ELECTRIC FURNACES

Design & Construction Of Electric Furnaces

Wilhelm Borchers

Translated By
Henry G. Solomon

ISBN 1-929148-48-8

Wexford College Press 2005

ISBN 1-929148-48-8

Wexford College Press 2005

TRANSLATOR'S PREFACE

THE present volume is an English version of the second German edition of "Die elektrischen Öfen" by Dr. Borchers, the well-known authority on electro-metallurgy. The English edition has been brought up to date as far as possible by the inclusion of descriptions of two of the most recent and most successful electric steel furnaces used on the Continent. These descriptions were also kindly supplied by Dr. Borchers, and have been added in an appendix, as the information concerning the furnaces was only obtained after the work had been translated and printed.

The method of treatment and classification of electric furnaces adopted is from the point of view of the heating system on which their action is based, and their development and application are examined in an authoritative and critical manner.

As close a translation of the original as possible has been adhered to, and the English equivalents of the metrical measurements given in the text have been added, together with extra references to English patents, the latter being enclosed in square brackets. For the purpose of an English book the subject-matter has been further arranged under suitable headings in chapters corresponding with the main sections of the German original, without, however, altering the sequence.

The recent rapid development, notably abroad, of the electric furnace is sufficient proof of the important part it is playing, and is destined to play in a still greater degree in the near future, in connection with all classes of metallurgical operations. Mention need only be made of the application of electric furnaces in the electro-metallurgy of steel and iron on the Continent, in America, and in Canada, in the aluminium industry in this country and abroad, and in the manufacture of the alkali metals, etc. By the aid of electric furnaces it should be possible to develop new industries, and in districts hitherto unsuitable for electrical enterprise, especially where the raw materials are readily obtainable for the production of the substance desired, and current can be cheaply generated and supplied, as by the utilisation of waste furnace gases and overhead transmission.

H. G. SOLOMON.

CONTENTS

CHAPTER I

INTRODUCTION AND METHODS OF HEATING

CHAPTER II

DIRECT RESISTANCE HEATING

FURNACES IN WHICH THE SUBSTANCE TO BE HEATED ITSELF FORMS THE HEATING RESISTANCE

CHAPTER III

INDIRECT RESISTANCE HEATING

FURNACES IN WHICH THE SUBSTANCE TO BE HEATED IS IN CONTACT WITH AN ELECTRICALLY HEATED SUBSTANCE

CHAPTER IV

DIRECT ARC HEATING

FURNACES IN WHICH THE SUBSTANCE TO BE HEATED FORMS ONE OR BOTH POLES OF AN ARC

CHAPTER V

COMBINED RESISTANCE AND DIRECT ARC HEATING

CHAPTER VI

INDIRECT ARC HEATING

FURNACES IN WHICH THE SUBSTANCE TO BE HEATED IS IN A SPACE HEATED BY INDEPENDENT ARCS: ELECTRIC RADIATION FURNACES

CHAPTER VII

FURNACES ARRANGED FOR DIFFERENT MODES OF HEATING

CHAPTER VIII

THE CONSTRUCTION OF ELECTRIC FURNACES

CHAPTER IX

THE APPLICATION OF ELECTRIC FURNACES

CHAPTER X

FURNACE OUTPUTS

APPENDIX

SOME RECENT DEVELOPMENTS IN STEEL FURNACES

CHAPTER I

INTRODUCTION AND METHODS OF HEATING

INTRODUCTION

Conversion of Electrical Energy into Heat.—Considering the facility with which electrical energy is converted into heat, the length of time over which our knowledge of this transformation of energy extends, as well as the advantages of this method of generating heat in solving the more difficult problems of technical heating, it is surprising that almost a generation has passed away since the invention of the dynamo before the construction of electric furnaces was more energetically developed and their field of usefulness increased. Even at the time of the publication of my first treatise on the development, construction, and working of electric furnaces in 1896, the number of furnace systems in practical operation was extremely small, with the exception of those used for experimental purposes.

Joule's Law.—In the conversion of electricity into heat we know that the heat developed must be equivalent to the work EC, if during the passage of a current of C amperes through a length of conductor the electromotive force consumed be E volts. As the heat equivalent is 0·24 calorie per watt-second, in t seconds the amount of heat produced, Q, expressed in gramme-calories, is—

$$Q = 0{\cdot}24E \,.\, C \,.\, t$$

If, instead of E or C, the resistance R (in ohms) be given, then either (E = CR)—

$$Q = 0{\cdot}24C^2 \,.\, R \,.\, t$$

or $\left(C = \frac{E}{R}\right)$

$$Q = 0{\cdot}24\frac{E^2}{R}\,.\,t$$

In conducting heat experiments, and in controlling the working of electric furnaces, the equation—

$$Q = 0{\cdot}24E\,.\,C\,.\,t$$

is evidently the most useful, as one always measures the drop in pressure across the terminals of the current leads of the electric furnace and, in addition, knows the value of the current. The three equations of Joule's law contain all the information desirable to enable the selection of the working conditions to be made when a definite heat effect is aimed at with a given resistance or a given current.

Degree of Heat in Different Conductors.—It is so self-evident that in conductors of the same material both the resistance and the heat produced increase with increasing length and decrease with increasing cross-section of the conductor, that it should hardly be necessary to mention these facts here.

It is equally clear that with differently conducting substances the cross-sections of the conductors must be increased and the lengths decreased for the production of equal quantities of heat with decreasing conductivity (increasing resistance).

Conductors having the same resistance naturally give the same quantities of heat for the same amounts of energy expended in the same interval of time ; the degrees of heat produced in the conductors by these quantities of heat can, however, give rise to different temperatures, not only in different substances having different specific heats, but even in one and the same conducting medium for differences in size of the resistance lengths.

Suppose we imagine, as an example for the first case, Fig. 1,

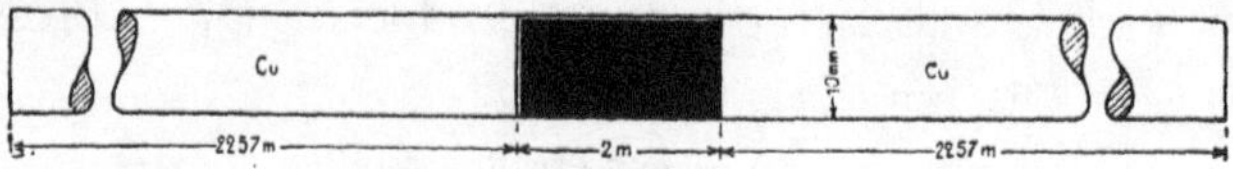

FIG. 1.

that a length of a copper conductor is replaced by a carbon rod of the same cross-section, and that the whole conductor has a uniform diameter of 10 mm., then the following result is obtained :—

	Copper.	Carbon.
Resistance per metre length . . .	0·0002215 ohm	0·5 ohm
Consequently, the length per ohm is .	4514 metres	2 metres
Weight per ohm in kilogrammes . .	3155 kg.	0·25 kg.
Specific heat	0·0952	0·2040
Temperature when heated with a current which in *t* seconds produces 100 kg.-calories per ohm . .	0·3° C.	1960° C.

As an example for the second case (Fig. 2), let us imagine

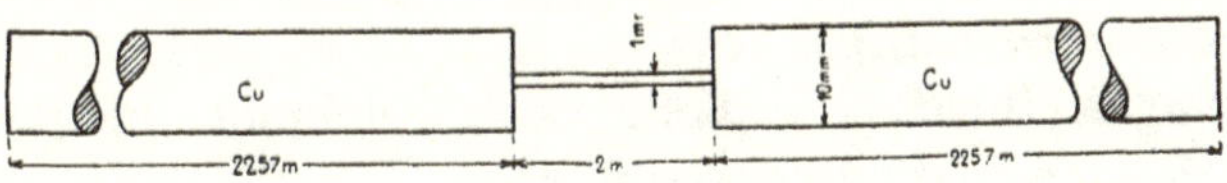

FIG. 2.

a thin copper wire of 1 mm. diameter connected to a copper conductor of 10 mm. diameter, so that each length of conductor has, as in the former case, a resistance of 1 ohm.

	Copper 10 mm.	Copper 1 mm.
Resistance per metre length . .	0·0002215 ohm	0·02215 ohm
Consequently, the length per ohm is	4514 metres	45·14 metres
Weight per ohm in kilogrammes .	3155 kg.	0·3155 kg.
Specific heat	0·0952	0·0952
Temperature when heated with a current which in *t* seconds produces 10 kg.-calories per ohm .	0·0332° C.	332° C.

Further examples are unnecessary here, especially as they do not give actual results, but only comparative values.

METHODS OF HEATING.

Choice of Resistance Materials.—It is clear from the above that in a given space we are able to concentrate almost any amount of heat in the shortest time. Even if we have to use as our resistance material a substance which is a good conductor, we can always choose it of such dimensions that it is raised to the desired temperature with comparatively weak currents. If, on the other hand, we are not tied to any definite material, then the selection of a suitable resistance for every case that can occur is mostly a very wide one. We are not even restricted to the state of aggregation of the materials at our disposal; solid, liquid, and gaseous substances can be employed. Finally, we can choose between metallic and electrolytic conductors for the resistance material.

Definitions of Resistance and Arc Heating.—Although it is in all cases a resistance that brings about the transformation of electrical energy into heat, nevertheless, one speaks of *resistance heating* in the narrower sense only, when the resistance is a solid or liquid; when gases or vapours form the resistance material, one speaks of *arc heating.*

Heating Methods.—In both cases different methods of carrying out the heating are possible:—

I. *Resistance Heating.*

1. The substance to be heated is itself connected to the electrical circuit as the conductor resistance—direct resistance heating.

2. The substance to be heated is in contact with an electrically heated resistance—indirect resistance heating.

II. *Arc Heating.*

1. The substance to be heated forms one or both poles of an electric arc—direct arc heating.

2. Combined resistance and arc heating.

3. The substance to be heated is situated in a space heated by the arc—indirect arc heating.

Before entering into the details of construction of furnaces, their development will be proceeded with in the separate chapters on heating.

CHAPTER II

DIRECT RESISTANCE HEATING

FURNACES IN WHICH THE SUBSTANCE TO BE HEATED ITSELF FORMS THE HEATING RESISTANCE

Davy's Experiment.—It was apparently Davy[1] who, in his experiments on the electrolytic decomposition of the earth alkali metals and earth metals, constructed the first electric furnace for direct and indirect heating, if only on the smallest scale. Although, in the electrolysis of molten alkaline hydrates, he accomplished the smelting through the agency of external sources of heat, in his experiments on *the electrolysis of aluminium oxide* he excluded any external supply of heat by adopting the following arrangement in a vessel filled with hydrogen gas.

A sheet of platinum was placed in communication with the positive pole of a voltaic pile comprising 1000 pairs of plates. This platinum sheet had on it a layer of alumina which had been moistened with water and firmly kneaded together. Into this mass, at the top, was introduced an iron wire connected to the negative pole of the above pile. The iron wire was soon raised to a white heat and melted at the place where it was in contact with the alumina. The metal mass, when cold, proved to be whiter and more brittle than iron. Treated with acids, a solution was obtained from which the alumina could be separated.

Pepys' Experiment.—Very soon afterwards, in 1815, we find an interesting report of an experiment, the somewhat unexpected result of which was the *cementation of iron wire* with diamond powder by the application of electric heating. A piece of iron wire, forming the resistance of an electrical circuit, was placed

[1] *Philosophical Transactions*, p. 16 (London, 1810).

in contact with some diamond powder; on passing through the current it became converted into steel by the electric heating. According to the *Philosophical Transactions* of the year 1815, Pepys[1] conducted this experiment as follows:—He bent a piece of pure, soft iron wire to an angle, and divided it longitudinally at the bend with a fine saw. Into the opening so formed he sprinkled diamond dust (Fig. 3), which he prevented from dropping out by means of finer wires. The part of the wire containing the diamond powder was also enveloped in thin leaves of talc. Thus prepared, the wire was connected to a battery. It very soon became red hot, and was kept at this temperature for six minutes. On opening the wire, Pepys found that the diamond dust had disappeared, and that the portion of the wire in which it had been enclosed was completely converted into steel.

FIG. 3.

Further information of the thickness of the wire, current density, etc., is not given; nevertheless, it is not only of historical but of practical interest, as it shows us that already, ninety years ago, the cementation of iron rods was effected by heating with a heavy current in contact with carbonaceous matter.

The Cowles Zinc Furnace.—Unfortunately, a fruitless attempt was made by the brothers Cowles,[2] in the year 1884, to apply

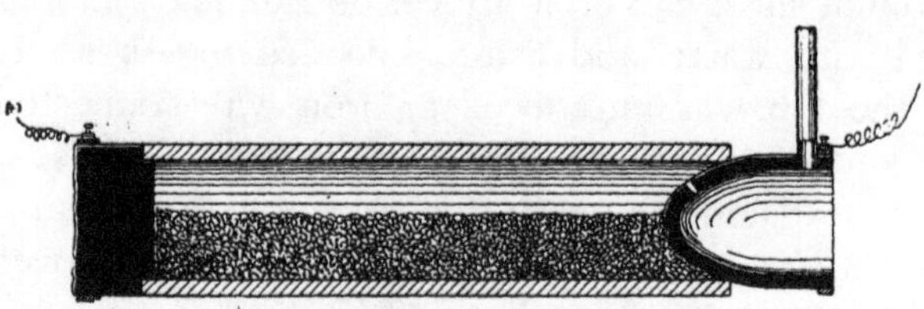

FIG. 4.

this heating principle to the *heating of the charge of zinc retorts* (Fig. 4).

The cylindrical retort, made of clay or other non-conducting

[1] *Philosophical Transactions*, p. 371 (London, 1815).

[2] [English Patent, 6994, June 9, 1885]; German Patent, 34,730, June 10, 1885.

material, is packed in bad heat conductors, and closed behind by means of a carbon plate connected to one of the poles of an electrical circuit. In the other end of the retort is inserted a closed graphite crucible, which communicates through an aperture with the interior of the retort, and is connected to the other pole of the source of electricity.

The ore which is to be reduced is mixed with bad-conducting material of high resistance—best of all, carbon used in electric lighting—and is completely enclosed by it. It is contained within the retort and becomes melted, volatilized by the glowing carbon particles on the passage of the current. The zinc vapours pass through the aperture into the crucible, where they condense, while the gases disengaged by the reaction escape from the crucible through an outlet pipe.

In this process the retort has not, as is usually the case, to withstand the greatest heat which is developed, and only a portion of which it has to transfer to the ore under treatment, but the maximum temperature is developed within the mixture of carbon and ore itself, and only a small portion of it is imparted to the retort.

In this case the whole arrangement of the apparatus was not happily chosen, for it must be borne in mind that the quantity of the conducting substance of the carbon present in the charge is constantly diminishing during the operation, while the conductivity of the retort is constantly increasing, first with the temperature, and then also with the accretion of conducting particles from the residues of reduction.

The Cowles Furnaces for Aluminium Alloys. – In Fig. 5 is shown diagrammatically a furnace in which liquid masses can collect inside the heating chamber. The vessel communicating with the furnace through the electrode on the left is to represent a condensing chamber through which the gaseous products of the smelting process are to be drawn off. In this case, also, the resistance between the poles of the electrical circuit is composed of the bad-conducting charge of aluminium oxide, which is mixed with reducing agents and any ingredients (copper, iron) required for the production of *aluminium alloys.* For the operation to be continuous, the charge should be replenished through feed hoppers in the roof of the furnace.

The further development of this furnace is given in the description, in *Industries* (1888), of the factory, shortly afterwards discontinued as unprofitable, of the Cowles Syndicate Company [1] for the production of aluminium alloys.

A 400-H.P. Crompton dynamo furnished a current of 5000 to 6000 amperes at a pressure of 60 volts.

The smelting furnaces consist of rectangular pits, the walls of

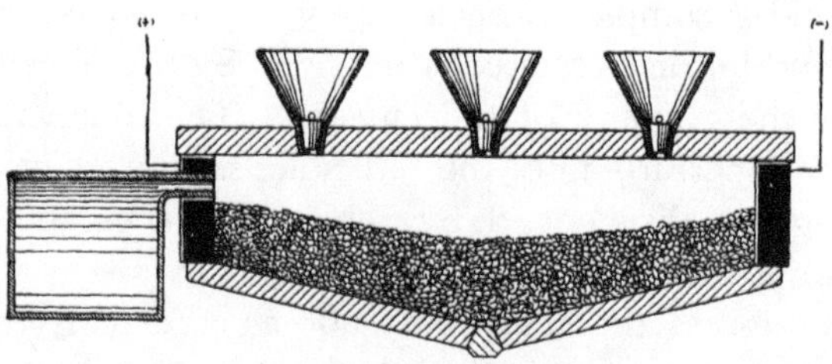

FIG. 5.

which are constructed of fire-clay. They are arranged together in a single row, but only one furnace is in operation at a time; while the others are being cooled, freshly charged, or emptied.

Two massive copper bars conduct the current, and run the whole length of the furnace room. They are arranged above the row of furnaces, the one in front and the other at the back. At the same time they serve as the running rails for the copper clamps provided with rollers. To these clamps flexible copper wire cables are connected, the lower ends of which are also held together in clamps. A suitable opening in the lower clamps enables the latter to be hung on correspondingly shaped copper rods. In this manner connection with the electrodes is established. Each electrode consists of a bundle of seven to nine carbon rods, each 64 mm. (2½ inches) in diameter, round which is cast a cylindrical head—of iron when ferro-aluminium is to be produced, or copper when aluminium-bronze is required. In the middle of the head is fixed one of the copper rods already mentioned. The electrodes are introduced through inclined cast-iron pipes in the opposite walls of the furnaces. By actuating a simple screw, the electrodes can be moved backwards and forwards as may be

[1] *Industries*, September 7, 1888, p. 237.

required for the regulation of the current. A layer of charcoal soaked with lime is first placed on the bottom of the furnace, and the electrodes are inserted. A sheet-iron frame is next placed in the furnace. Within the frame the space is filled with charcoal, ore, and metal, and the space between the frame and the furnace walls with charcoal; the frame is then withdrawn. Some fragments of retort-carbon are now thrown into the furnace to establish a bridge for the passage of the current between the electrodes. Any space still remaining empty is then filled up with charcoal, and, finally, a cover with holes in the centre is put on. The gases are ignited as they escape through the opening in the cover, and are conveyed through a pipe into a chamber in which the alumina, which has been carried over, subsides. Through a tapping hole in the bottom of the furnace the alloy accumulating there is run off. The slag, consisting of a very intimate mixture of alloy and charcoal, is broken up, and the charcoal is separated by washing with water. The alloy obtained in this way is then added to a fresh charge.

In the furnaces of this plant the daily output consisted of 750 to 1000 kg. (1653–2205 lb.) of ferro-aluminium, or aluminium-bronze, containing 15 to 17 per cent. of aluminium. By remelting with addition of copper, the bronze was obtained in the commercial varieties having 1·25, 2·5, 5, 7·5, and 10 per cent. of aluminium, and was cast into ingots weighing 5 to 6 kg. (11–13 lb.). The expenditure of electrical energy amounted on the average to 50 horse-power-hours per kilogramme of aluminium (22·7 horse-power-hours per lb.).

Fig. 6 gives a view of the furnace room, Fig. 7 represents a longitudinal, and Fig. 8 a transverse section of a single furnace. E, E are the electrodes, each consisting of nine carbon rods, about 30 mm. ($1\frac{3}{16}$ inch) in thickness, round which are cast the cylindrical metal blocks M. In each of these metal blocks a copper rod K is let in on the opposite side to the carbons. The whole arrangement is movable within the pipe R, in which it can be pushed backwards or forwards by the screw S. The current connections are made by means of the stranded copper cables L, which are clamped in the connecting pieces V, also of copper, and with these can be hung on the conical ends of the rods K, which project through the pipes R. The rods K are guided by the iron guide-blocks F. The screw works in the collar Z, fixed to K.

FIG. 6.

It is manifest, from Fig. 7, which shows the present arrangement of the carbon rods inside the furnace under working conditions, that these carbon rods, especially the middle ones

FIG. 7.

projecting beyond the bundle, which in the descriptions are, really, wrongly termed electrodes, only act as resistances in completing the circuit of a strong source of current. It is they which are first of all heated and impart their heat to the mixture packed round them. Gradually the carbon particles present in the mixture form resistance paths for the electric current, while the rods E become partially consumed by the oxygen of the metal oxides.

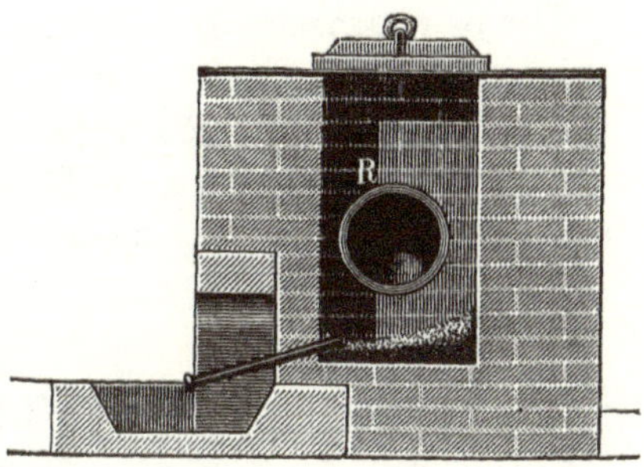

FIG. 8.

The Héroult Aluminium Furnace.—In the electric smelting industry the Héroult furnace has acquired very special importance. It was developed from the experimental apparatus shown in the accompanying illustration. In the latter, a smelting crucible stands on a conducting plate. The space between the crucible

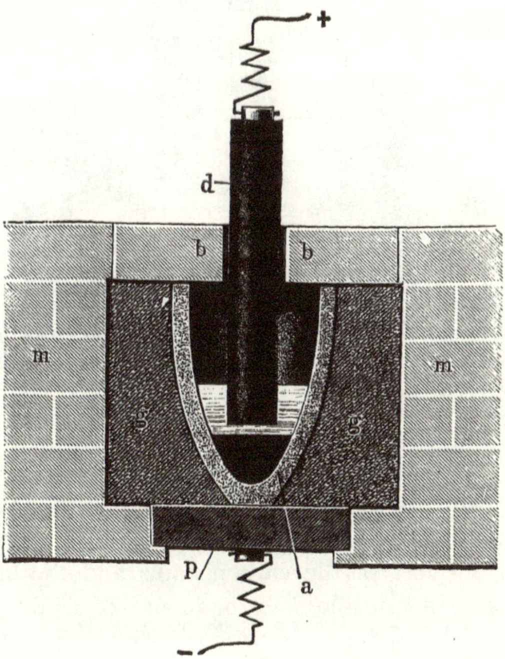

FIG. 9.

and the masonry is filled up with carbon powder. The carbon anode passes through the cover, and the crucible itself forms the cathode (Fig. 9).[1]

The first furnace[2] intended for practical working was arranged as depicted in Fig. 10. An iron or other metal box, insulated from below and provided with a heavy lining of carbon plates, serves as the smelting vessel. The carbon plates were held together by means of a carbonaceous cement, which consisted of tar, molasses, or glucose. In order to obtain a very favourable conductance by means of the most intimate contact of the outer surfaces of the carbon walls and the inside of the casing, the carbon crucible was made use of as the core in the manufacture of the casing as an iron or other metal casting.

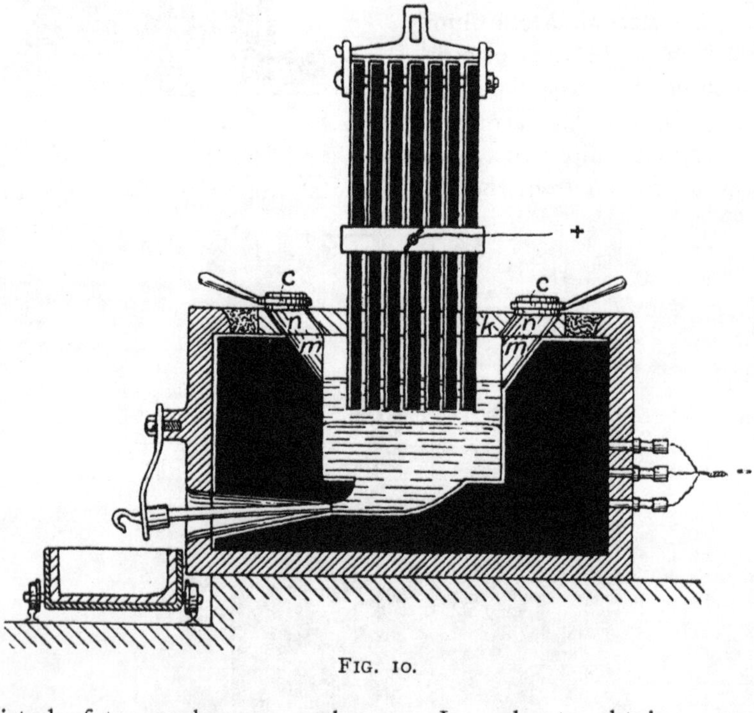

Fig. 10.

[1] United States Patent, No. 387,876, August 14, 1888.

[2] German Patent, No. 47,165, December 8, 1887; English Patent, No. 7426, May 21, 1887, not, however, quite the same as the German Patent.—[H.G.S.]

This small pit-furnace was connected to the negative wire of the circuit.

Carbon blocks, or bundles of carbon rods or plates, served as the anode. With the exception of an aperture necessary for the vertical movement of the carbon anode, the furnace was covered with graphite plates *k*, in which a few small openings, *n*, were made for the introduction of material. Corresponding to these openings, *n*, which are kept closed during the working by the movable covers *c*, the necessary apertures, *m*, are provided in the side walls of the basin. These channels, *m*, *n*, also serve to carry off the gases disengaged within the furnace. The space between the graphite cover and the rim of the casing is filled in with charcoal powder.

The operation is started by introducing copper, preferably in the broken state, into the crucible; the anode is then lowered into contact with the copper, which becomes melted by the current. As soon as the bath, now acting as the cathode, consists of liquid copper, alumina is thrown in and the anode slightly raised. In the now melting alumina electrolysis at once sets in, with the separation of the aluminium in the fused copper, *i.e.* with the formation of aluminium-bronze.

At first current densities of 23,000 to 25,000 amperes per sq. metre (14·84–16·13 amperes per sq. inch) were worked with, requiring 10 to 15 volts. These high-current densities, which to-day are no longer employed, were probably rendered necessary because the bath worked with consisted mainly of aluminium oxide admixed with only a small amount of cryolite, or other flux; whereas to-day one chiefly employs cryolite, etc., as the solvent and flux for the aluminium oxide, added at regular intervals to the bath according to the strength of the current.

The earlier method of working, which required a higher temperature to be maintained than that of to-day, further necessitated a metal bath on the carbon lining to receive the aluminium, as without this metallic solvent a large part of the aluminium would have combined with the carbon lining to form aluminium carbide. In the construction of the furnace, the main defects of working at that time were not to be looked for; *Héroult had, in fact, made in this way the first electric furnace which possessed in the fullest degree the advantages of the pit-furnaces so extensively employed on account of their suitability for metallurgical work; it was the first electric furnace which, like the pit-furnace, could be kept in continuous operation, the charge of which could be introduced from above, and the products of which could be periodically tapped from the bottom of the smelting hearth.*

And in spite of the defects inherent to the first methods of working, Héroult achieved still another considerable success with his process. As we see from the arrangement of the anode, which dips deep into the electrolyte, he prevented the formation of an arc on melting the aluminium compounds to be electrolyzed. His idea, to electrolyze fused aluminium compounds free from water with such high current densities that the electrolyzing current at the same time generated the heat for melting both the electrolyte and the metal separating out on the cathode, made it possible to practically carry out a process which, at the time of the appearance of Héroult as electro-metallurgist, was already completely known for forty years in all other respects. *His process is the most perfect method of heating for effecting electrolysis in the fused electrolyte.*

FIG. 11. Scale 1/10.

Borchers' Aluminium Furnace.—The removal of the defects existing in the first method of carrying out the Héroult process was attempted in another way by altering the apparatus. As in my laboratory experiments during the years 1888 to 1890 I had vainly endeavoured to obtain pure aluminium by the electrolysis of molten aluminium oxide in carbon vessels, I constructed an experimental furnace in which intentionally, as far as I know, *the principle was for the first time laid down, that the walls of the electric furnace should be constructed of the same materials of which the furnace charge consisted, and that these walls should be preserved by external cooling.* At first I lined the metal casing of the furnace with aluminium oxide, later with cryolite. In the former case the cooling of the furnace walls by the surrounding air sufficed; in the other case with small furnaces water cooling had at least to be employed in order to preserve the cryolite wall during the working. As the latter construction embodies all the parts of the first, I will only describe this one:—

A sheet-iron case serves to support a base of burned clay or metal, on which a copper cooling vessel stands, and in this a compressed wall of powdered cryolite. The cathode, a carbon block fitted into a holder which can be cooled, is passed through the base-plate. The adjustable carbon anode is suspended in the crucible from above, and naturally passes through the cover when one is used. To start the furnace, the anode is brought into contact with the cathode. It is then quickly drawn back until an arc is struck, and some cryolite is strewn over the floor of the crucible by means of a spoon, where the salt, especially in the neighbourhood of the arc, speedily fuses. As soon as the cathode is covered with molten metal, the furnace can be fairly quickly filled with cryolite, and at the same time the anode is gradually raised until it is a few centimetres away from the cathode. Aluminium oxide is then added to the electrolyte as the aluminium separates out.

The Minet Aluminium Furnace.—Minet also evolved several resistance furnaces in his experiments and practical work on the extraction of aluminium during the years 1887 to 1894. The first furnace was furnished with external heating, which at once excludes its application [1] for the object mentioned. Of the other two types of furnaces the second is a Héroult furnace, a metal vessel lined with carbon and having carbon electrodes suspended in it from above. Finally, the third furnace, like the first, could

[1] Cp. Borchers, "Electrometallurgie," 3rd edition, 1902, p. 140; ["Electric Smelting and Refining," Dr. W. Borchers and Walter G. McMillan, 2nd English edition, p. 141].

probably never fulfil its purpose for continuous working (Fig. 12). The cathode hangs near the anode in the molten mass, which is again contained in a carbon-lined metal vessel. He recommended the furnace for the production of very pure aluminium. As electrolytically there is no connection whatever between the crucible wall and the anode, the molten mass should remain free from iron and other metallic impurities, especially as the metallic walling of the crucible can, of course, be cooled in such a manner that the temperature of the inner side remains lower than that of the molten mass, so that the latter cannot penetrate up to the crucible walls. But how about the efficiency of the furnace? The metal flows down the cathode, and collects in a basin hollowed out in the centre of the furnace floor. From here the metal can be withdrawn through an outlet channel. The use of this furnace can only be considered for the separation of metals of high solution-pressure in fused electrolytes, and such a metal should, in this case, be constantly flowing down the anode in a thin stream, and drip from the end of the same. A considerable quantity of these metal drops dissolves in the electrolyte, into which free metalloids also diffuse from the anode; for the path from the electrode to the floor of the smelting hearth should not be too short, as otherwise the metal layer on the floor becomes connected to the electrical circuit as a very active intermediate electrode.

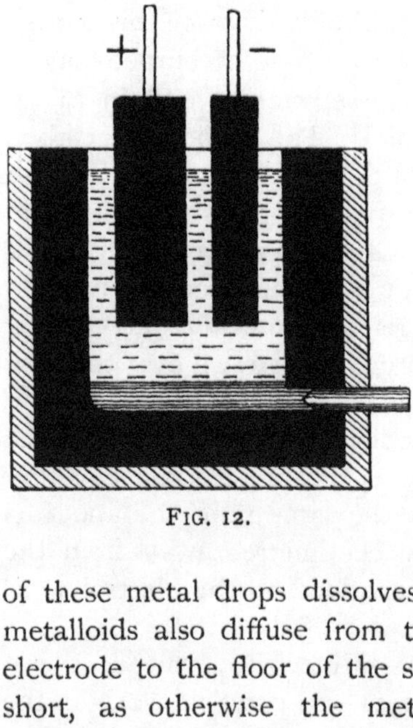

FIG. 12.

The Héroult Furnace.—That in the course of time the Héroult furnace should undergo more or less radical changes in its details of construction one can readily understand from the advantages and the far-reaching application of this method of heating.

The high-current densities with which one had to work at the start, as long as fluxes rich in aluminium oxide were used, brought many evils in their train. The quantity and the pressure of the gases which collected on the comparatively small anode surface (oxygen and the products of combustion arising from it) formed an almost unbroken gas envelope through which the electric current could only pass by forming an arc, if only a very short one. The result was a specially strong development of heat in the environment of the anode, still further augmenting the gas pressure and also the gas layer, with, of course, a corresponding increase in the electromotive force and the working costs. Further damage was suffered through the consumption of the electrode, which naturally increased with the temperature of the electrolyte in connection with the tip of the anode dipping into it. I do not mean here the unavoidable consumption of the anode which takes place with the evolution of oxygen on that part of the carbon which dips into the electrolyte, but rather the consumption of the electrode above the electrolyte. In addition, the atmospheric air enters here, and with all the more speed the hotter the parts of the carbon become which are situated immediately above the electrolyte in consequence of the upward transfer of the heat. In quite a short time we can see how a rapidly deepening cavity forms round the whole electrode, how also in this already endangered region the cross-section of the electrode diminishes, and thus the resistance increases at this point; this again occasions a further rise in the temperature, and, consequently, an increase in the rate of admission of the air and consumption of the substance of the anode. In a few hours thick anode blocks are so eaten into above the bath that a slight knock suffices to break off the considerably thicker anode-head still hanging in the bath. In Fig. 13 I give a photographic view of the anode of a small apparatus.

At first the attempt was made to remove all these defects by the use of rotary devices. By rotating the smelting vessel, or the electrode, and thus continuously displacing the electrolyte against the anode, it was thought to establish a better contact between the anode and the electrolyte; it certainly succeeded, now and again, in partially breaking down the gas envelope on the surface of the anode, thereby reducing the injurious overheating; but the first really effective remedy consisted in the reduction of the current density. Of course, many other causes are instrumental in preventing the intimate contact of the electrolyte with the substance of the anode. Thus particles of the electrolyte having low boiling-points, apart from any gaseous electrolytic products disengaged on the anode, can produce the *Leidenfrost* phenomenon on the anode surfaces, whereas, on the other hand, molten substances with large surface tension prevent the wetting of the anode, much after the style of mercury. Rules cannot, therefore, be established for all cases, how far the current density may be taken on the anode in carrying out the electrolysis in the molten flux, for even carbons of different manufacture exhibit a varying behaviour in this respect.

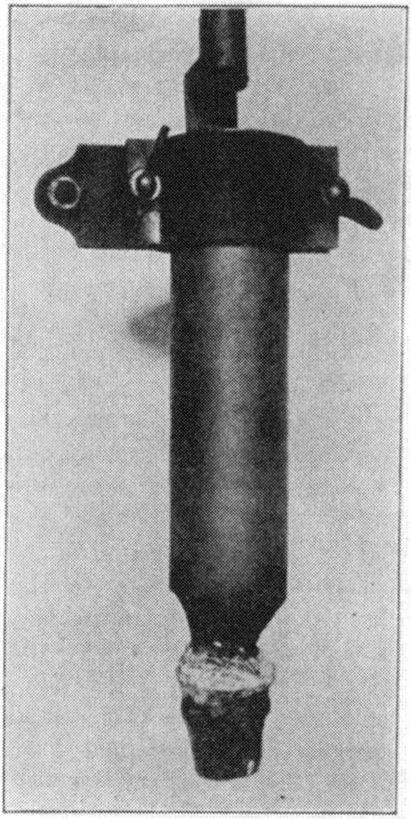

FIG. 13.

The reduction of the current density, especially in the aluminium industry, was at once obtained by increasing the quantity of the flux or solvent for the aluminium oxide, and correspondingly reducing the quantity of the latter, which, of course, has in all cases to be added during the operation in proportion to the strength of the current used.

By using easily fusible baths that was achieved which appeared to me impossible in the experiments carried out on the small scale, namely, the extraction of the metal free from carbon, *i.e.* the prevention of the formation of metal carbides in spite of the use of electrolytic vessels having a carbon lining at least on the floor. So already in the year 1890 the technical press could report the solution of the problem of the extraction of pure aluminium in smelting vessels lined with carbon, *i.e.* in the original *Héroult furnaces.* According to *Industries*,[1] the furnaces depicted in the annexed illustration (Fig. 14) had been at that time used at Neuhausen and Froges for the extraction of pure aluminium, when even metal electrodes were temporarily used as the cathode. The crucible was first charged with cryolite, and later, at regular intervals, with aluminium oxide in proportion to the quantity of the aluminium separated out.

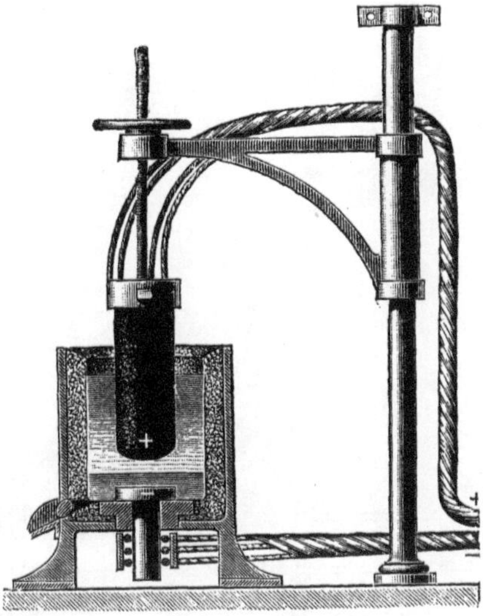

FIG. 14.

The Hall Furnace of the Pittsburgh Reduction Company.—An almost exact copy of the original *Héroult furnace* is the furnace used by the Pittsburgh Reduction Company for the manufacture of aluminium on the Hall process, as Fig. 15 shows on the following page. As is known, Hall obtained a number of patents for easily fusible, good-conducting electrolytes of specific gravities, such as appeared desirable for the subsidence of the specifically light aluminium.

Bradley's Patents.—The simplest furnace of this class is without doubt that of Bradley. Although the patents which were granted in the December of 1891, and in the February and April of 1892, had already been applied for in the year 1883, nevertheless, the furnace had only become generally known through patent processes in the United States in the year 1896. Bradley in these patent specifications himself states that the main object of his invention "is to dispense with the external supply of heat for the object of smelting ores." "In order to obtain this result," says Bradley further, "a current of greater strength, or intensity, should be used than is necessary for electrolysis alone, and in this manner the ore is obtained in the molten state by the heat

[1] *Industries*, vol. **8**, p. 499 (London, 1890).

which is formed on the passage of the current through the molten mass." The electric current has, therefore, to perform two different functions. The one consists in maintaining the

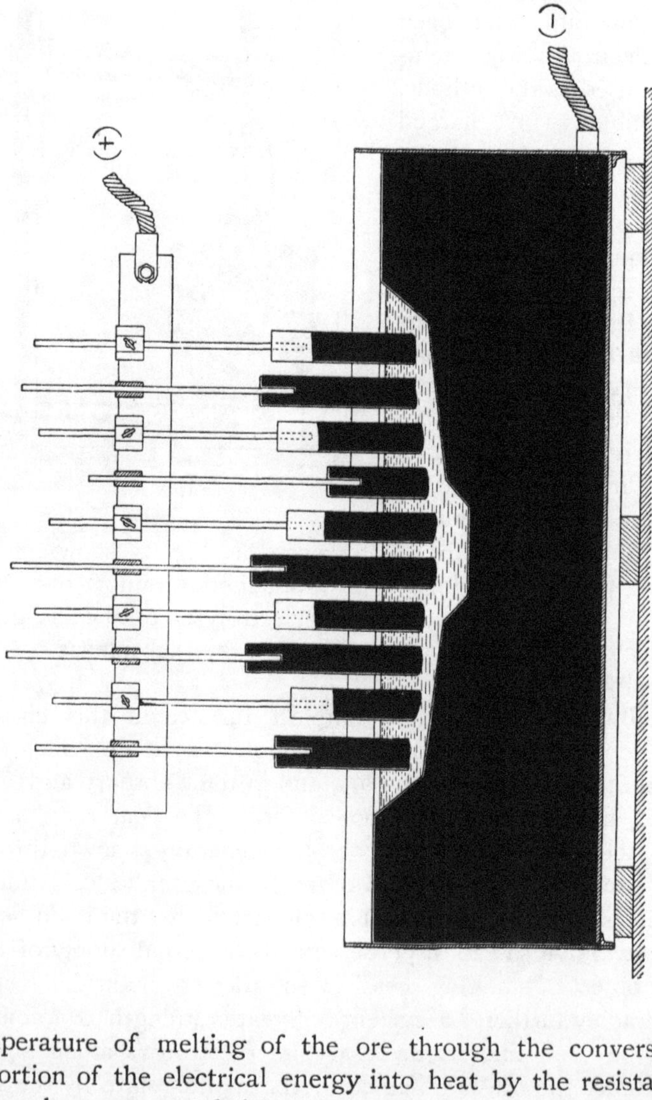

Fig. 15.

temperature of melting of the ore through the conversion of a portion of the electrical energy into heat by the resistance of the molten mass, and just at the place at which the heat for melting is necessary. The other function consists in bringing about the desired electrolytic decomposition.[1]

Another feature of the invention consists in the suppression of the customary crucibles for the smelting vessel, by making the latter out of heaps of the ore itself which is to be reduced. As this is as little destroyed by the action of the flux as by the liberated gases, continuous working can be obtained by simply replenishing with fresh ore from the outside or from the side walls of the heap at the rate at which the reduction goes on.

To carry out the work, a heap of more or less coarsely pulverized ore is piled up on a suitable hearth and hollowed out for the reception of the flux. The first reduction of the ore is effected by bringing together the ends of two electrodes connected to the poles of the dynamo, or other suitable source of current, and then separating them again sufficiently far from each other to start the arc. If now the ore be thrown into the furnace, it will soon become liquefied by the heat of the electric arc and form a conducting electrolyte. The arc will, of course, at once disappear as soon as a conducting liquid, the molten ore, exists between the electrode, as *the current is now conducted by the flux, in which heat is generated as in incandescent lamps.* The arc is,

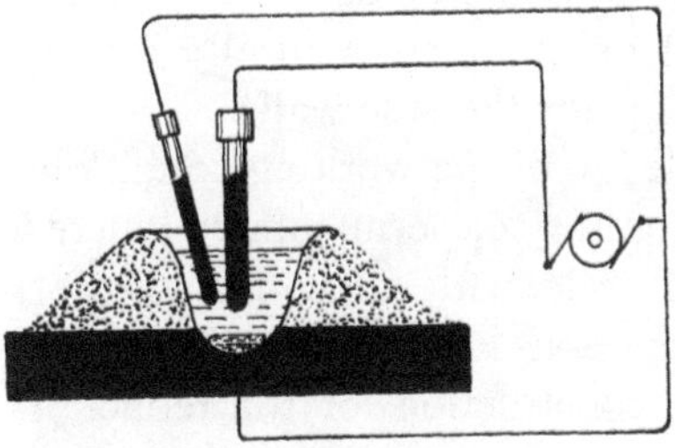

FIG. 16.

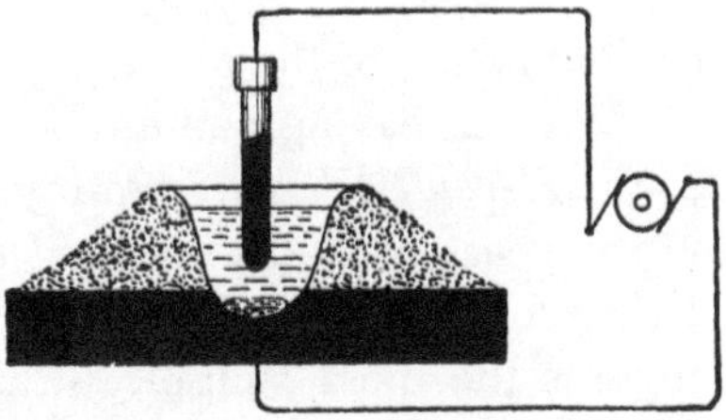

FIG. 17.

therefore, only employed at the commencement of the process for melting-down purposes, while the heat required for smelting the ore is maintained by so-called "incandescence." As long as the ore is kept in the molten state, the aluminium is deposited at the cathode ; at the anode fluorine gas is liberated.

[1] United States Patent, No. 464,933, granted December 8, 1891 ; No. 468,148, February 2, 1892 ; and No. 473,866, April 12, 1892.

As soon as the operation has been properly started, the electrodes must be moved further away from one another, so that the metal separating out does not cause any short circuits and is not attacked by the fluorine.

The figures given above (16 and 17) show various arrangements of the electrodes, in which, however, a carbon plate, which forms the floor of the furnace, is always used as the cathode.

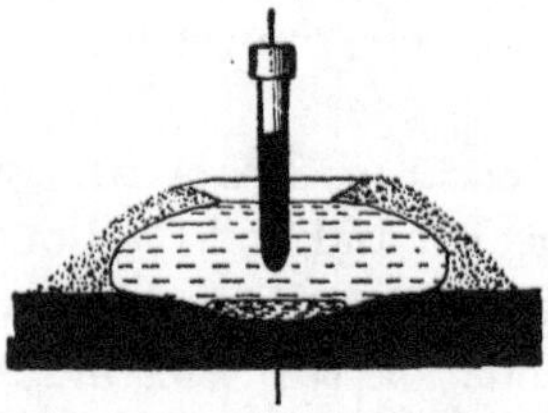

FIG. 18.

For working on a large scale, this apparatus is unfortunately unsuitable. The flux does not preserve the form shown in the patent specification. In consequence of the cooling action of the air, the rim of the bath solidifies close up to the top electrode, while beneath the dam, which is a poor heat conductor, the flux expands laterally and easily breaks completely through it (Fig. 18).

Furnaces of the Héroult Works.—According to Winteler's book on aluminium (Braunschweig, 1903), who, judging from the information he gives of arrangements of various works operating on the *Héroult* principle, appears to be well informed as regards their working, the furnaces used to-day at these works consist of iron vats, having a carbon floor lining and a layer of solidified cryolite, formed or obtained naturally by the cooling action of the air, as the lining for the side walls.

Without having had any knowledge of the working conditions of the *Héroult Works*, from my laboratory experiments, conducted in the years 1888–92, on the small scale, with a few horse-power, I therefore arrived at the same results as the Héroult Works as regards the most suitable furnace construction for the reduction of aluminium.

Winteler gives the following description of these furnaces, together with drawings of a bath for 3200 amperes :—

On the large scale a simple wrought-iron tank is preferably used, suitably stiffened with angle iron as a protection against alterations in shape. Cast iron should not be employed, because it easily springs with sharp variations in the temperature, and because, in addition, its melting-point is approximately at the same temperature as that of the flux. The iron tank is protected against the corrosive action of the bath flux by means of a solidified layer of the flux itself, so that actually the material of the molten mass itself forms the walling of the vessel, and the iron construction only gives this the necessary stiffness.

The floor of the bath serves as the anode. It is lined with carbon plates, on which collects the metal which has separated out. When the floor is once covered with the latter, then the metal itself functions as the anode. The floor is covered with a layer of carbon plates, because iron alloys slightly with aluminium, and because, at the start of the operation, through resistance heating, local temperatures are set up which might melt the iron.

The following are suitable dimensions of the bath for a current strength of 3200 amperes: Length = 1050 mm., breadth = 550 mm., height = 300 mm. (3 ft. $5\frac{3}{8}$ in. × 1 ft. $9\frac{5}{8}$ in. × 11 ft. $\frac{3}{16}$ in.). The annexed drawings (Figs. 19, 20, 21) show a normal aluminium bath based on these proportions.

The bath is best erected on insulated supports above an air-shaft, which is provided with a draught regulator. The current of air streaming past the floor of the bath cools it to a temperature at which the aluminium that has separated out is still in the liquid state, but which is below that at which aluminium can alloy with iron in larger quantities. After the bath has been in operation for some time, the carbon layer on the floor gets gradually destroyed, and the molten metal then comes into direct contact with the iron floor. Now, aluminium has the property of not fluxing with iron at temperatures which are not materially above its melting-point, whereas at higher temperatures it easily attacks it and alloys with it. This property is taken account of here.

An arrangement for regulating the air draught is also advantageous, because the radiation losses of the bath at different times of the year are so very varied. In the summer the cooling must be made stronger, in the winter less strong.

The cathode conductor-bars can be screwed direct to the bath.

The conductors to the anode, to which the copper rods of the carbon electrodes are fastened by means of screw terminals, are arranged immediately above the bath. Their height above the latter depends on the length of the electrodes. They consist of copper bus-bars resting on insulated supports and sufficiently strong to carry the various electrodes.

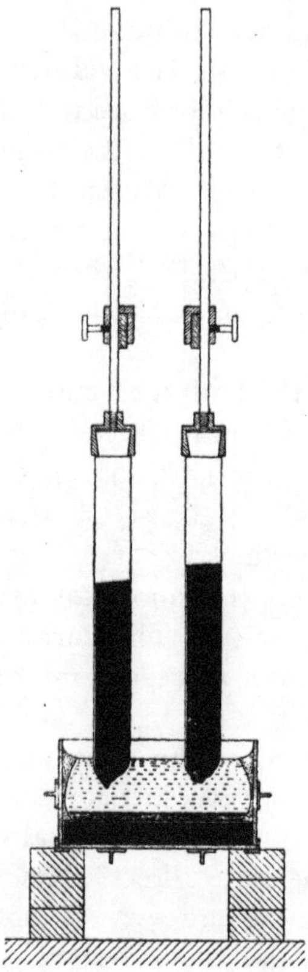

FIG. 19. Scale $\frac{1}{25}$.

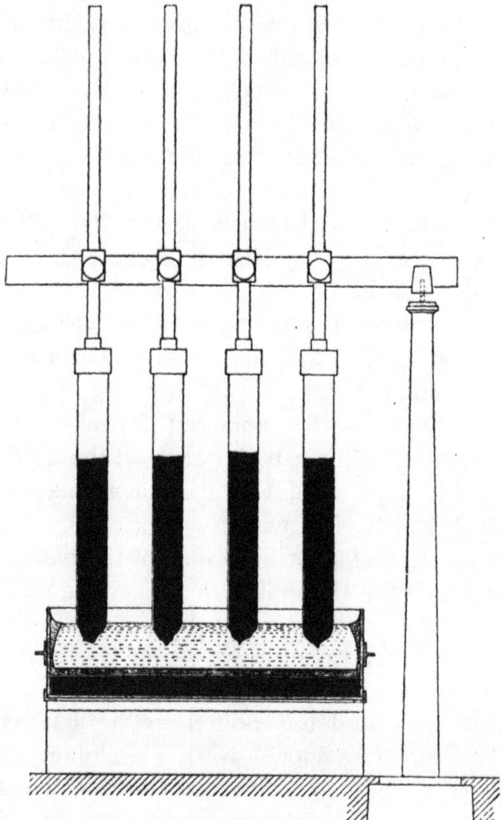

FIG. 20. Scale $\frac{1}{25}$.

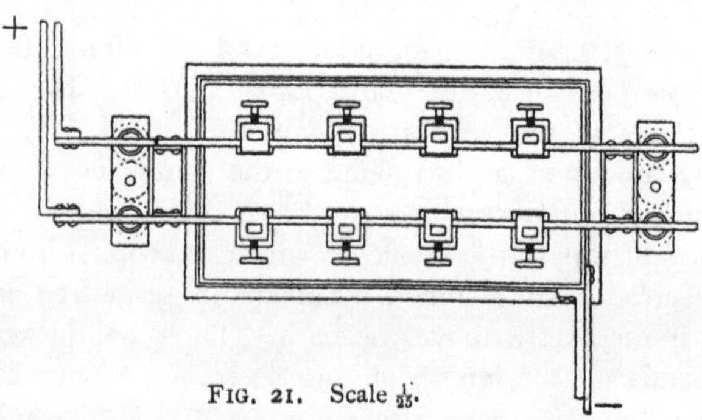

FIG. 21. Scale $\frac{1}{25}$.

Current is led to the latter in the following manner: A flat copper rod, provided with a quick thread at its lower end, is screwed direct into a corresponding hole in the upper part of the carbon. This arrangement is a very suitable one, as the carbons can in this manner be consumed for the greater part of their length with only a small amount of waste, as is shown in the accompanying figure (Fig. 22). Another method is to dove-tail the one end of the carbons, and then to screw the copper conductor into an iron contact-piece which is carried by the dove-tailed end. With this arrangement, however, the carbon waste is considerably greater than in the first-mentioned case. Further, the contact leaves much to be desired, as the voltage loss between the copper and the carbon amounts to about 0·6 to 0·8 volt.

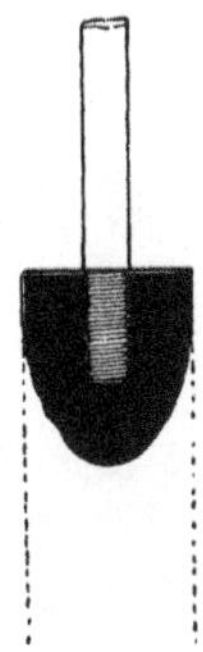
FIG. 22.

It is unnecessary to provide a means of producing a fine regulation of the electrodes, as the distance between the anode and cathode need only be approximately kept, and is best about 6 cm. A regulation is, in fact, only undertaken at intervals of about two hours, provided that the bath is in normal working condition. With a bad method of working, however, a more frequent regulation may certainly become necessary. This happens when the carbons, in consequence of faulty setting, burn away irregularly; also when, with the bath incorrectly put together, undissolved sandy masses accumulate on the floor, necessitating a more frequent raising of the electrodes.

The Héroult Furnaces of the British Aluminium Company.—In the Héroult Works,[1] at the Falls of Foyers, in Scotland, the work is carried out in iron smelting vessels, having carbon-lined floors of a clear cross-section of about 1500 mm. × 750 mm. (4 ft. 11 in. × 2 ft. 5½ in.) with currents of 8000 amperes, corresponding, in round numbers, to a current density of 7000 amperes per square metre (650 amp. per square foot).

The Urbanitzky and Fellner Aluminium Furnace.—A furnace construction by Urbanitzky and Fellner,[2] as far as it is intended

[1] *Journ. of the Soc. of Chem. Ind.*, 1898, p. 308.

[2] German Patent, No. 82,164, January 29, 1895; English Patent, No. 7265 (1895).

for the manufacture of aluminium, according to the patent specifications, is equipped in an unnecessarily lavish manner with cooling arrangements, and, indeed, in places where they are applied the least, while the pit walls are too thick and have to

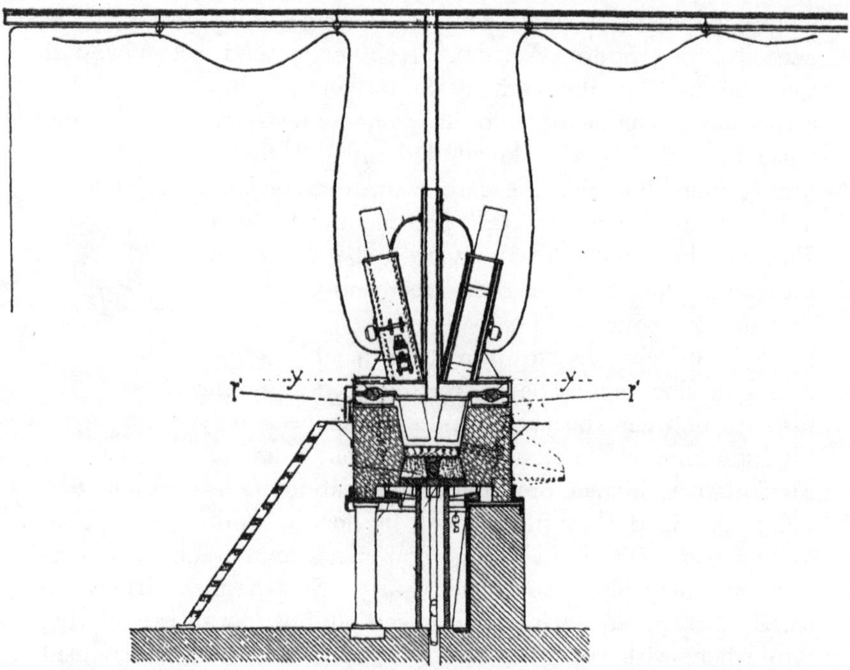

FIG. 23.

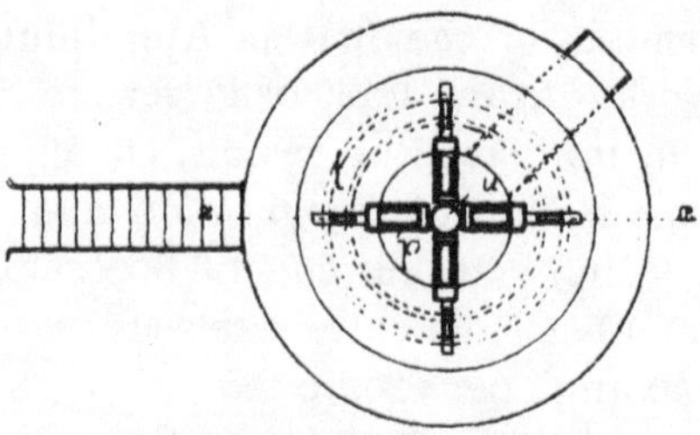

FIG. 24.

forego the cooling mostly applied here. The short pit, built up of basic, non-conducting material, is closed underneath by a wedge-shaped carbon or metal body, which can be cooled.

The positive electrodes consist of carbon bars arranged edgewise and secured by their guides to a cover which rests on rollers situated in a groove in the supporting plate. The latter is arranged at the mouth of the pit, and in such a manner that the cover can be easily turned round its central axis. The cover, as well as its supporting plate, and the anode-guides are made hollow and are traversed by a constant stream of cooling water. In the middle of the cover is a vertical pipe, which serves to carry off the evolved gases as well as for the introduction of the alumina into the furnace; this pipe is also surrounded with a cooling jacket.

The furnace is also suitable for working with the electric arc; it was recommended later by Urbanitzky[1] for iron and steel smelting processes, like his first pit-furnace.

Resistance Furnaces for the Extraction and Refining of Iron.—Towards the middle of the last decade of the past century we see the first resistance furnaces of this kind being established for the extraction and refining of iron since the small experiment of Pepys conducted in 1815.

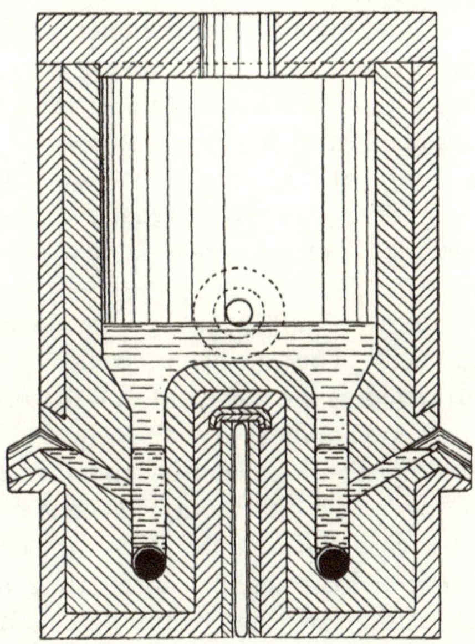

FIG. 25.

The de Laval Steel Furnace.—In the de Laval furnace,[2] for which patents were already applied in the year 1892, the smelting chamber consisted of a low pit-furnace, divided, like the fire-bridges of reverberatory furnaces, into two parts by means of a water-cooled, central division-wall of non-conducting material. In each of the narrow cavities so formed were the conductor-rods (coloured black in Fig. 25) for connecting the electrolyte to the

[1] *Zeitsch. f. Electroch.*, 1895, **2**, p. 350.

[2] German Patent, No. 80,462, June 12, 1892; [English Patent, No. 15.793, March 8, 1892].

current supply. Above these the metal to be refined collected in the molten state, while the actual heating resistance was composed of an oxide salt, or other appropriate fusible substance. For iron, magnetite was proposed. The metal, previously reduced from ores below its melting-point, was thrown through an opening in the furnace cover into the smelting bath, from which a refining action was expected, as in the Martin process, and in which the pure metal dissolved. To prevent electrolytic action, which was not intended here, heating with alternating current was recommended. For running off the metal and regulating the total height of the liquid, so-called syphoning channels in the side-walls of the furnace were employed; while for running off the fused ore, which acts as the resistance material and gradually becomes contaminated with impurities, and for regulating its height above the bridge, a scum hole at the level of the liquid was provided.

The Taussig Steel Furnace.—The Taussig furnace,[1] according to all that was known about it from patent specifications and reports in the years 1893 and 1894, represents a thoroughly well-designed smelting apparatus from a constructional standpoint, although at that time it did not achieve commercial success. It is, however, noteworthy that the principles of construction of this furnace were ten years later again used for the same purpose (Gin).

Figs. 26 and 27 give a sectional elevation and plan of the furnace. The metal electrodes presenting large surfaces are arranged at the front side of the branches of a horseshoe-shaped melting chamber, so that, instead of facing each other, they are side by side. Current is conducted to and from the electrodes by rods screwed into them, and shut off by stuffing-boxes from the iron case surrounding the furnace space. To these rods are connected the actual leads coming from the source of electricity. From the smelting chamber a channel, usually shut off by a slide from the smelting chamber, leads to a mould arranged beneath the furnace. The slide is moved from without by means of a screw-threaded hand-wheel through a double-armed lever and a screw spindle passing air-tight through the cover of the case.

[1] English Patent, No. 3573 (1893); German Patent, No. 77,125, August 31, 1893.

With the slide open the liquid contents of the furnace chamber pass through the above channel, which widens within the furnace

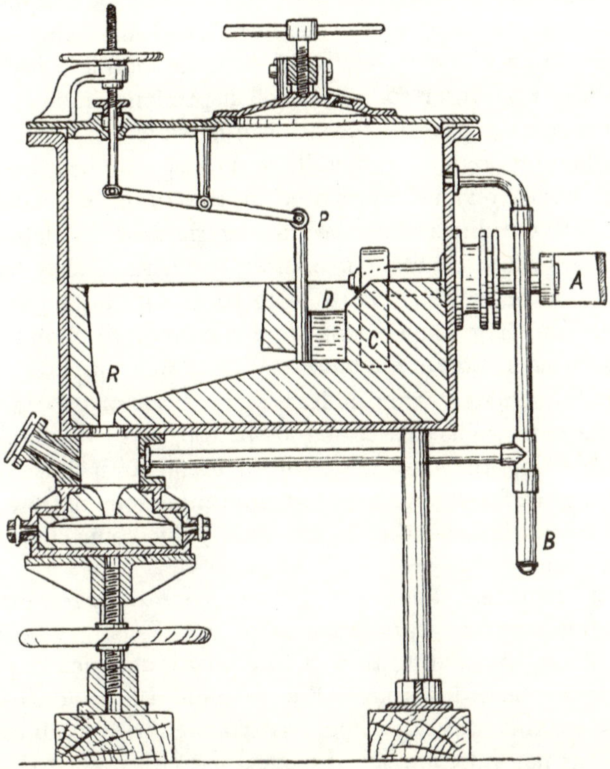

FIG. 26.

to a trough, and then through a pipe situated in the floor into the mould. The latter rests on a plate, which can be moved in a vertical direction by a right- and left-handed screw rotated by a handle, the one end of the screw engaging with a nave, and the other end with a stationary nut. In this manner an air-tight joint is made between the mould and the furnace, which also permits of

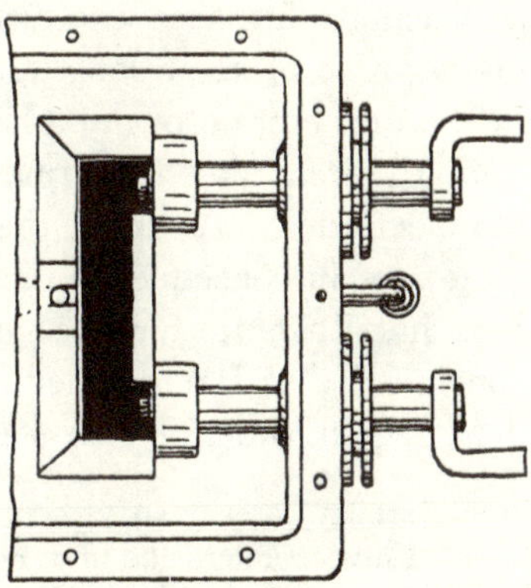

FIG. 27.

the attachment of an inspection window as well as of the connection of air suction pipes. The latter are connected to the furnace near the smelting chamber and to the pipe leading to the mould. The furnace charge is introduced into the smelting chamber through the cover, which can be hermetically closed. In the cover is inserted a second inspection window, through which the operation in the furnace can be watched.

Crushed ore, oxide, or metalliferous slag can be thrown into a metal bath previously prepared in the furnace, or it can be subjected to the direct action of the electric current. The trough within the furnace chamber can, after the molten slag has been run off from the melting hearth, also serve for the separation of the metal and slag when castings are not going to be made.

Wikström's Steel Furnace.—Wikström [1] proposed to pass pig-iron immediately after it leaves the blast or cupola furnace through a narrow channel closed at the top, and to quickly refine it by subjecting the narrow stream of metal, as it flows through, to the action of a lateral air-blast from numerous narrow twyers, at the same time heating it electrically by connecting it to a heavy-current circuit.

Gin's Furnace.—The hearth-furnace, repeatedly recommended by Gin for several years, inclines to the Taussig construction, but has a longer smelting trough, to the ends of which the current circuit is connected through the medium of cooled cast-steel contacts, in this way avoiding any contact with carbon. This furnace has not yet come into practical use (Figs. 28, 29, 30, 31).

Borchers' Furnace.—That the extended pit-furnaces, known as Rachette furnaces, can also be arranged for electrical heating, on account of their long narrow hearths, I already pointed out in the first edition of my "Electric Furnaces," [2] at the same time giving there a sketch for one of the possible methods of arranging the electrodes. As after the publication of my book, one of the reviewers, in consequence of this reference, and probably from the fact that Rachette furnaces are used in the lead-smelting industry, concluded that I recommended electric heating for smelting lead ores, I wish, therefore, to state distinctly that I did not express,

[1] German Patent, No. 76,606, November 24, 1893.

[2] "Entwickelung, Bau und Betrieb elektrischer Öfen," 1st edition, 1896, pp. 36–39. Published by W. Knapp, Halle (Saale).

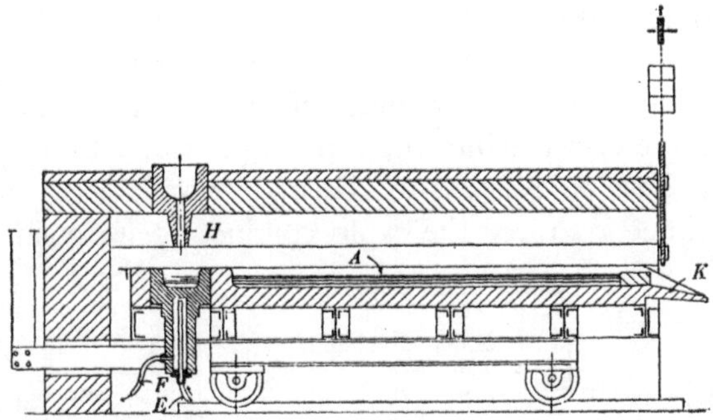

FIG. 28.

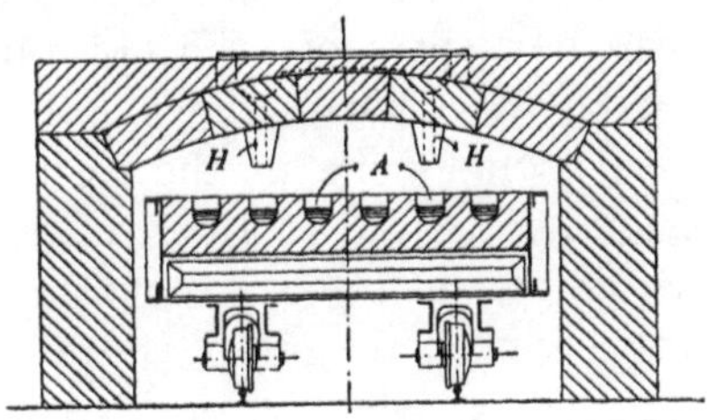

FIG. 29.

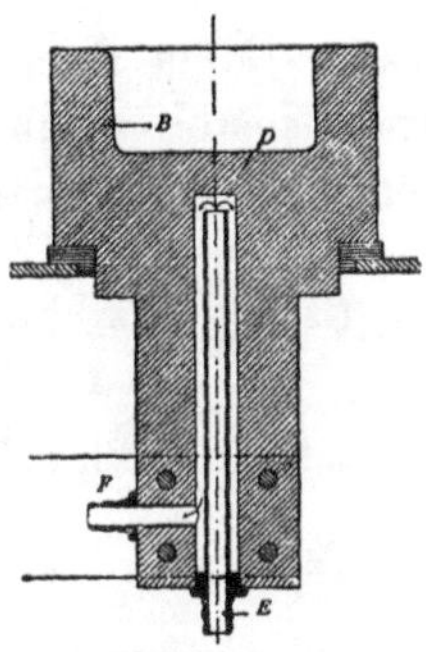

FIG. 30.

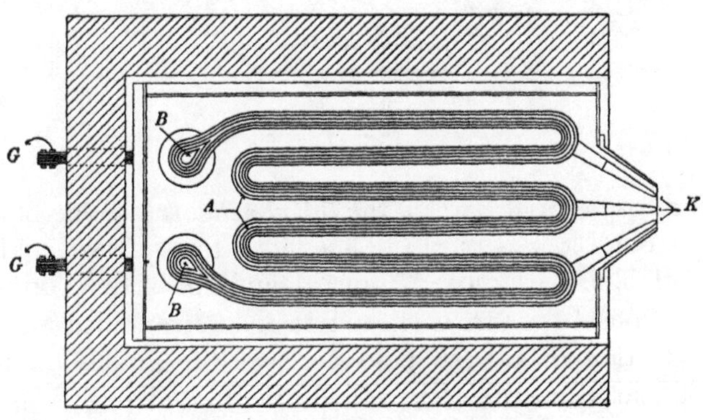

FIG. 31.

and, naturally, did not entertain, any such idea. Figs. 32 and 33, of which the former is a half-sectional elevation and the latter a ground plan, show the method of introducing the proposed carbon-block electrodes into the furnace, as well as the method of connecting these pole-blocks to the main circuit. By means of a suitable connection-piece the cable conductors are pushed on the

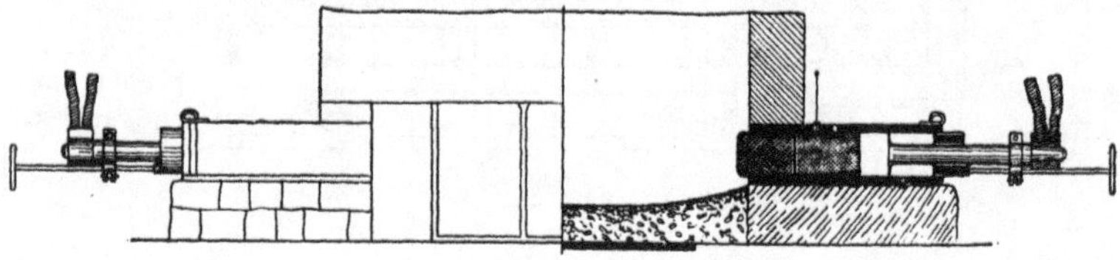

FIG. 32.

one end of a thick copper rod, the other end of which is inserted into a piston arranged so that it can be kept cool. The piston, and with it the carbon blocks, can be moved with the aid of a hand-wheel and screw-spindle. As soon as one of the carbon blocks has been consumed, the piston has arrived at a part of the guide-box where, between the last carbon block and the piston, a slide can be let down. When this happens the piston is drawn

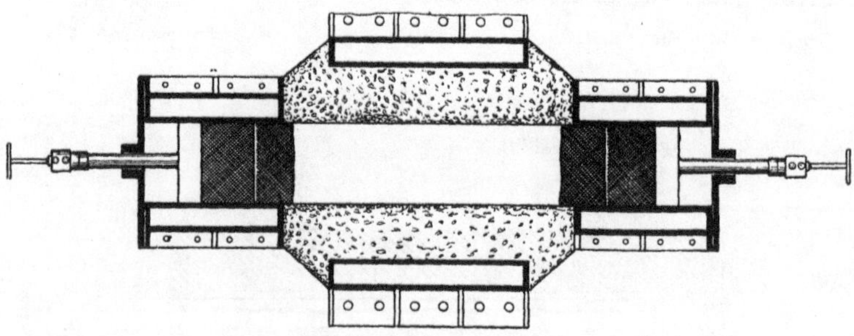

FIG. 33.

back, as also the front part of the lid closing the guide-box, and a fresh carbon block is put in. The lid is then closed again, the piston and block are moved forward, and the slide is drawn up. The operation of the furnace need not be interrupted at all during the introduction of a fresh pole-piece, as the guide-box for all these pole-parts consists of iron, and therefore allows the passage of the current when the piston has been drawn back. The hearth is, naturally, best made of the same oxide as it is intended to reduce, by pounding coarse lumps mixed with finer powder between the hearth walls, which are built up of stones and iron. For the protection of the mass of the hearth and the pole-bodies, the side walls of the hearth and of the pole guide-boxes are made hollow so that they can be kept cool. The pit walls can, of course, be built up of ordinary masonry.

Carbide Furnace of the Société des Carbures Métalliques.—Quite noteworthy is a furnace design by the Société des Carbures

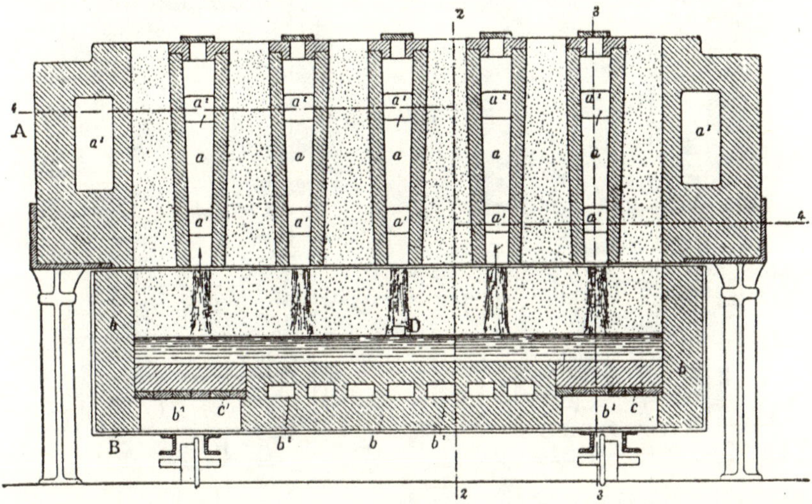

FIG. 34.

Métalliques,[1] in which the molten products on the hearth of an extended pit-furnace also serve as the heating resistance.

The stationary furnace pit is divided by a number of gas outlet channels, *a*, into a number, greater than unity, of feed hoppers, in which the charge is warmed in advance by the gases as they burn in the channels *a*.

In Figs. 34–36 a^1 are the openings for the admission of air into the channels *a*. The ignited gases are led through the holes a^2 into the side channels a^3, in which they circulate round the furnace and finally escape through an exhaust flue, not shown in the drawing.

[1] German Patent, No. 101,832 of November 23, 1897.

D

It should be observed that these outlet channels narrow slightly towards the lower end to facilitate the falling of the mass to be melted. B is the movable part of the furnace which contains the carbide bath and the two electrodes, c, c^1. The latter are arranged horizontally; they can, however, be slightly inclined. The walls are composed of a non-conducting material which has to be capable of withstanding the temperature required for melting the carbide. Below the electrodes, c, c^1, hollow spaces are

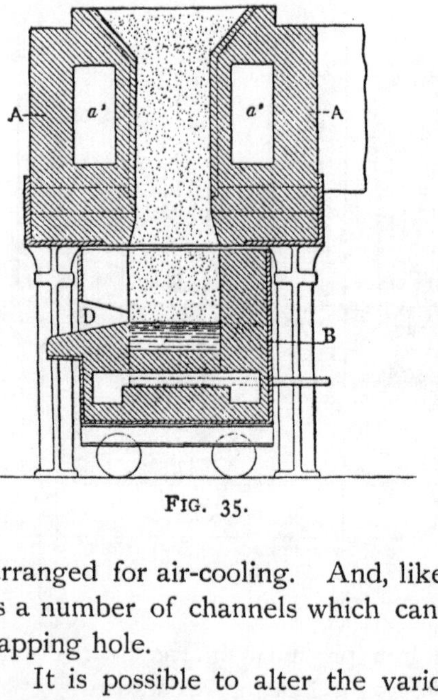

FIG. 35.

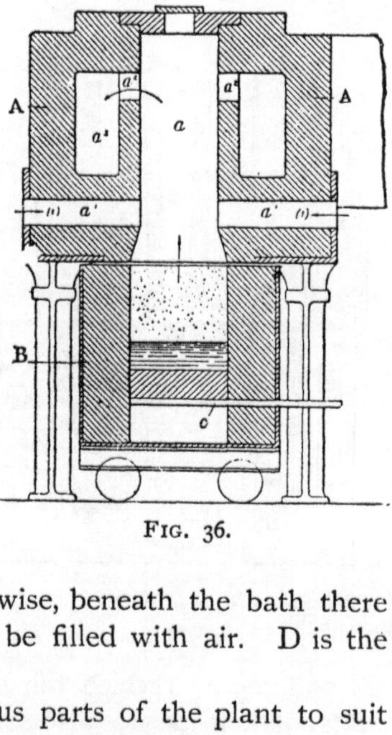

FIG. 36.

arranged for air-cooling. And, likewise, beneath the bath there is a number of channels which can be filled with air. D is the tapping hole.

It is possible to alter the various parts of the plant to suit different requirements. For instance, it is not absolutely necessary to arrange the furnace floor so that it is movable, as, like the upper part, it can be made stationary. Instead of the outlet channels, simple dividing walls can be used to split up the furnace charge.

Ruthenberg's Furnace.—A simple furnace construction is indicated by Ruthenberg[1] in an American patent. The electrode holders simultaneously serve to support the small smelting shaft-furnace, and are arranged for cooling. The molten ore overflows

[1] United States Patent, 647,614 of April 17, 1900.

into a portable pot from a hearth erected beneath the shaft, but independent of it (Figs. 37, 38, 39).

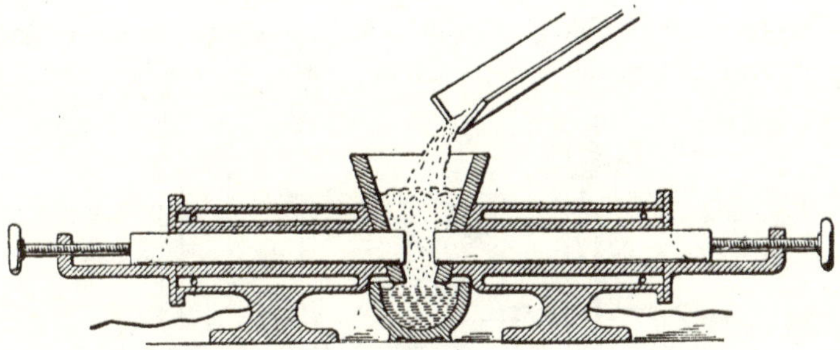

FIG. 37.

FIG. 38.

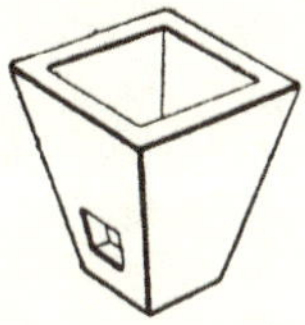

FIG. 39.

Carbide Furnace of the Compagnie Electro-Metallurgique des Procédés Gin & Leleux.—In the furnaces made by the former Compagnie Electro-Metallurgique des Procédés Gin & Leleux [1] (Figs. 40–43) for the manufacture of carbide, which were shown at work at the Paris Exhibition of 1900, the electrodes, or connections for the mass to be heated, were comprised on the one hand of the floor of the portable melting vessel, and on the other hand of a carbon pole suspended in the crucible from above. The arrangement of both contacts is of importance for the method of working and the economical result. The crucible floor is lined with two layers of carbon; the bottom layer, in contact with the iron floor of the crucible itself, is of superior conductivity, and carries a second layer which does not conduct so well. The object of this arrangement is to protect the molten mass on the floor from cooling, and subsequent solidification by

[1] Borchers, "Die Elektrochemie auf der Pariser Weltausstellung," 1900. W. Knappe, Halle a. d. S.

the heat generated electrically in the upper floor-layer. The top electrode is made up of four rods of high conductivity enclosed within carbon material of inferior conductance.

The furnace is charged from above, and the molten mass is tapped and the gases escape through the channels arranged in

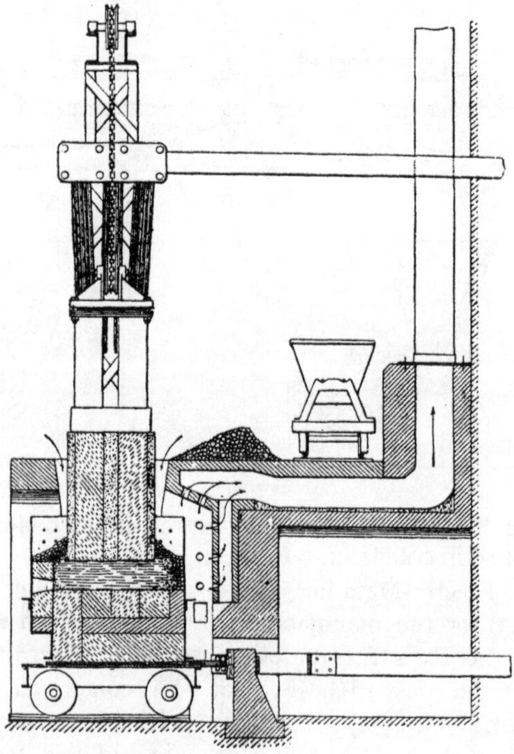

FIG. 40.

the back wall, so that the workmen are protected from every inconvenience and danger due to carbonic oxide and dust. After some time crusts of solidified carbide, nevertheless, accumulate in the retort, which naturally disturb the working. In order to be able to easily remove these residues, the retort, as already pointed out, is once for all mounted on a truck ; in addition its top casing must be lifted off, and the residues, fused into blocks, can then be readily removed. Apart from the top electrode, down which the furnace charge slides and up which the hot gases ascend and act both mechanically as well as chemically through the presence of carbonic acid, the remaining parts of the furnace require but

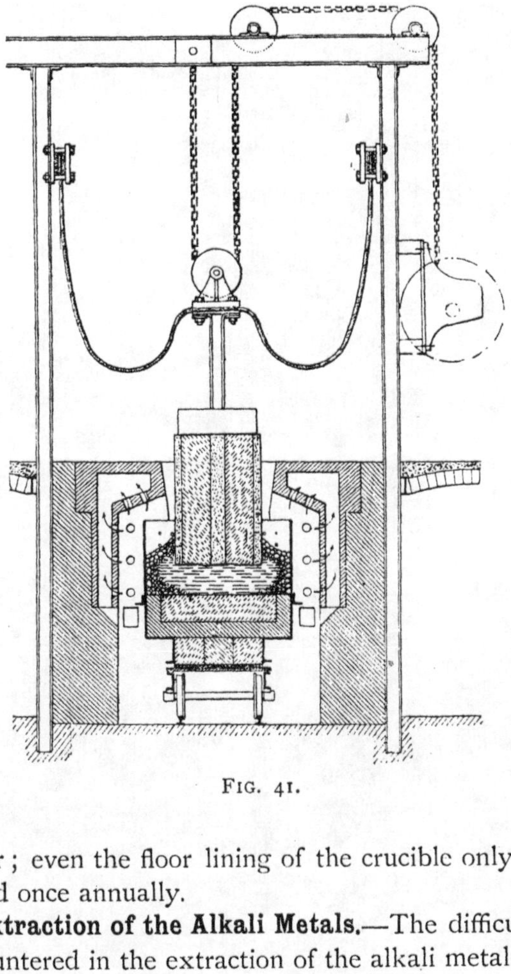

FIG. 41.

little repair; even the floor lining of the crucible only requires to be renewed once annually.

The Extraction of the Alkali Metals.—The difficulties which were encountered in the extraction of the alkali metals, especially sodium, by the electrolysis of molten alkaline chlorides and other halogen salts, as is well known, led to the application of easily fusible metals, such as lead, tin, and others, for the cathode, for the purpose of absorbing the electrolytically-separated alkali metal and to prevent, or at least delay, its re-solution. Although one had to work with such high current densities in the electrolysis of molten alkaline salts that even the more difficultly fusible salts are kept liquid by the heat produced by the current, the first experimental apparatus were still equipped with external heating,

FIG. 42.

so that they cannot be regarded as purely electrical furnaces. Some designs of furnaces, in which the heat required for smelting is entirely produced electrically by the resistance method, were patented by Ashcroft[1] in the year 1903, for the extraction of

[1] English Patent, No. 12,377 (1903).

alkali-lead alloys. In these constructions the inventor, however, completely ignored the experiences which had hitherto been gained in the electrolysis of this alloy, so that they by no means answered their purpose, even disregarding their otherwise faulty

FIG. 43.

construction. While in the electrolysis of fused halogen salts, the anode is mostly arranged in the middle of the smelting vessel, in order on the one hand to be able to run off the halogen through a jacket surrounding the anode, and on the other hand to be able to utilize the smelting vessel as the cathode, in cases in which a very high current density of the cathode is required for the decomposition of a metal, the cathode is situated in the middle of the vessel.

In the apparatus built by *Castner and Becker* for the extraction of alkali metals an annular anode embracing the *centrally situated cathode* is provided within a wide smelting vessel, whereas in the apparatus constructed later by me for *Stockem* for the extraction of alkali and earth alkali metals I used the *melting vessel* itself *as the anode.*

The Castner Alkali Furnace. – Castner's apparatus [1] consists of an iron melting vessel A (Fig. 44), into which the cathode H

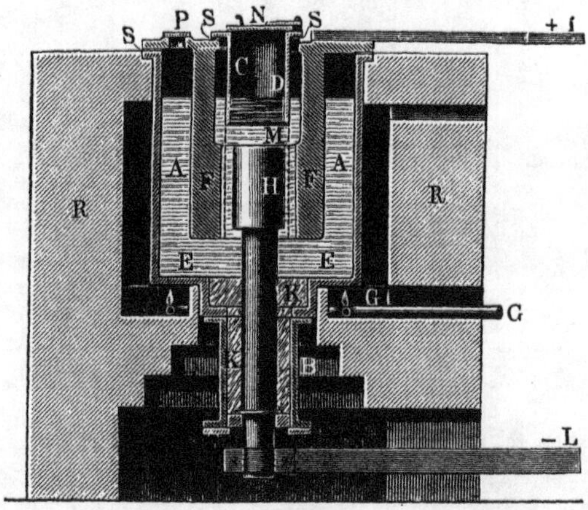

FIG. 44.

is introduced through the floor. For making the cathode air-tight and securing it, the lower and slightly narrower part of the vessel and a pipe B joined to it are filled up with acid alkali K before the retort is set to work; the acid alkali K solidifies after a short time. The anode F dips into the molten acid alkali E, kept liquid by gasburners G. The anode is fixed to the lid, and in such electrolytes can, of course, also be composed of metal. Between H and F hangs a cylinder M of wire gauze as a diaphragm. To this is attached the collector pipe C at the top, in which accumulate the hydrogen and metal D dissociated from

[1] German Patent, No. 58,121; English Patent, No. 13,356 (1890).

the oxygen, the latter escaping through the opening P in the cover. The pipe C is covered by the lid N, which fits sufficiently loosely to allow the hydrogen to pass through. For scooping out the sodium, Castner used perforated ladles in which the metal remains in consequence of its surface tension, while the caustic soda runs through. The various parts of the apparatus are insulated from one another by asbestos plates, S. I and L denote the current leads.

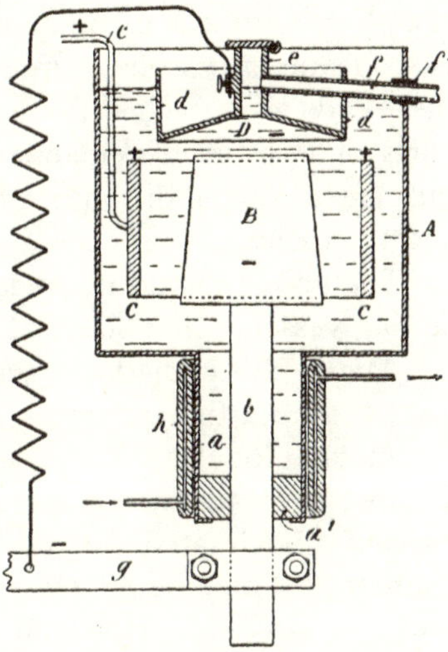

FIG. 45.

The Castner-Becker Furnace.—The Castner apparatus was altered later by Becker (Fig. 45) in the following manner.[1]

In the floor of a metal vessel A a wide pipe *a* is inserted, through which the rod *b* passes, and to the end of the latter is attached the actual cathode B. The lower end of the pipe *a* is sealed by a ring, *a'*, of lava, porcelain, fire-proof stone, or other such material. The rod *b* passes through the centre of this ring *a'*. The pipe *a* is surrounded by a double-walled shell *h*, or some other suitable cooling arrangement. The object of cooling the pipe *a* is to make the electrolyte contained in it pasty, or, if possible, to harden it, so that the electrolyte between the pipe *a* and the seal *a'* cannot escape. The cathode B consists of a metal piece, or, if necessary, of retort-carbon, and is easily made conical to enable the metal globules which form on the surface of this cathode to easily mount up vertically on the surface; it can, of course, be made in any shape suitable for this purpose.

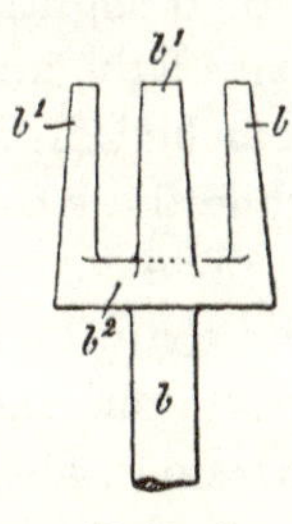

FIG. 46.

Fig. 46 shows, for example, a special form for the cathode; in this case it consists of a number of rectangular, square, or round

[1] German Patent, No. 104,955 of January 21, 1899; [English Patent, No. 11,678 of June 5, 1899].

rods, b^1, cast on the metal piece b^2. The latter is fixed to the rod b. With this arrangement the surface of the cathode can be considerably increased, and thereby the deposited metal concentrated on a smaller superficial area at the surface of the bath. The annular anode C envelopes the cathode completely; it can consist of one or several pieces. According to the ingredients of the electrolyte it is made of retort-carbon or metal. This anode is suspended from one or more rods *c* which serve as conductors; these rods are attached to the outer surface of the anode, or to parts of it, in such a manner that between these rods and the cathode a longer distance exists than between the anode itself and the cathode.

The anode should not reach down to the bottom of the containing vessel A, but stand at most at the height of the cathode.

Above the cathode is suspended a metal cone D which is insulated from the apparatus, and the object of which is to collect the metal globules which rise to the surface of the electrolyte. To its rim is attached a vertical or nearly vertical crown, *d*. In its centre is situated a vertical, thick water-pipe *e* of fairly large diameter. The pipe *e* is closed at the top by a heavy cover, which acts as a valve, or is closed in some other manner, and is provided with a slightly inclined outflow pipe *f* which leads through the wall of the containing vessel. Where this pipe *f* passes through the wall of A it is insulated by a ring f^1 of asbestos, porcelain, or other suitable material. The diameter of the metal cone D should slightly exceed that of the cathode B and be smaller than that of the anode C, so that all the metal globules which become detached from the cathode get beneath this conical collector, while the gases liberated at the anode can flow past it.

The conical collector D must not dip deeper into the electrolyte than the height of the rim *d*, so that the electrolyte does not cover the metal cone D. The surface of the metal cone is, therefore, always in contact with the air, whereby any excessive increase of heat is avoided. When the electrolyte with which one is working is very hot, the metal cone D can be cooled by directing on it a current of cool air, or by letting water drip on it, which is at once vaporized, or by using any other cooling method.

Fig. 47 shows, for example, a conical collector, the screen of which is formed of a hollow wall through which a gaseous or liquid cooling stream is passed. The object of cooling the collector cone is to lower the temperature sufficiently so that the free metal does not volatilize and also does not become oxidized in flowing out of the discharge pipe *f*.

The collector cone is joined to a branch circuit of the negative lead *g* through a resistance which only allows a very small fraction of the main current to pass through it. In this way the cone acts as a subsidiary cathode, so that the metal which is set free on the main cathode B and rises into the collector cone again becomes negative, while it passes along the inner surface of the cone and rises in the outlet pipe. The distance from the main cathode B to the subsidiary cathode D is traversed so quickly by the metal that it is unable to again dissolve in the electrolyte. As soon, then, as it comes into contact with this subsidiary cathode, the metal forms a part of the latter, so that it is no longer attacked by the bath.

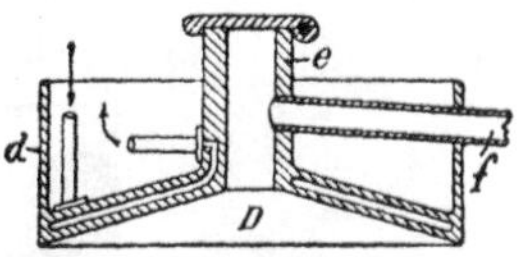

FIG. 47.

The shape of the collector cone is only mentioned as an example; any other type suitable for the same purpose can, of course, be employed.

When the gases liberated at the anode are to be collected, it is only necessary to close the apparatus with a lid having an outlet pipe.

The containing vessel A is erected either on a cast base or on masonry. In the vessel A *the electrolyte is kept fused by the heat developed by the passing current*, whereby only the central portion requires to be maintained in the liquid state, which ensures a long working life to the apparatus.

As the metal which flows out of the small pipe *f* is brought to a moderate temperature by the application of one of the above-described methods of cooling, it can, if not too strongly exposed to oxidation, be collected in any suitably-shaped moulding flask placed beneath the outflow. If, however, the metal be easily oxidized, an almost air-tight flask is used which is provided at the top with an opening through which the end of the outflow pipe *f* is inserted.

Borchers' Furnaces for Alkali and Earth Alkali Metals.—If, in the electrolysis of fused alkaline hydrates, special iron bodies were preferably suspended in the iron melting vessels to act as anodes, so in other cases the simultaneous application of the smelting vessel as anode may be desirable. For the *electrolytic working of fused alkali and earth alkali halogen salts* melting vessels of carbon were naturally desirable. As the manufacturing

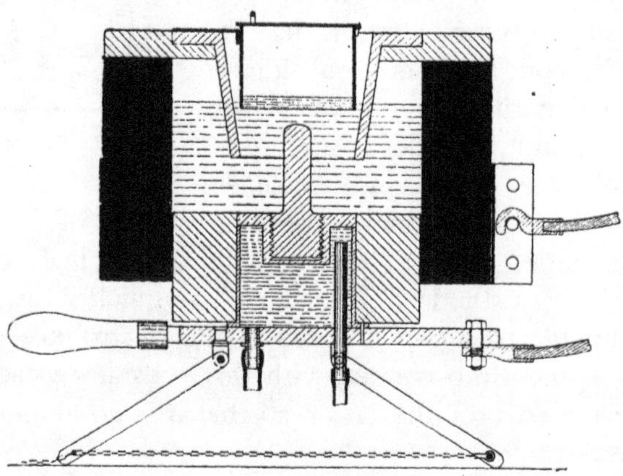

FIG. 48.

cost of such vessels of even moderate dimensions was very high, I had them constructed of uniform bars which could be easily compressed (Figs. 48, 49, 50).

The melting vessel consists of a cylinder of carbon bars, or segments, joined and held together by a metal ring, like the staves of a cask. It serves as the anode during the process of working, and it is connected to the supply circuit through the metal ring. The cylinder is closed at the bottom by a cooling vessel, which, for the sake of simplicity, serves at the same time as the cathode-holder. The cathode itself consists of an iron bar placed in the cooling vessel. By means of an insulating body of clay the anode is shielded from the cooling. The joints between the clay ring and the carbon segments were filled in with asbestos. The lower part of the apparatus was stamped out with desiccated sodium chloride, which, in consequence of its melting-point being higher than the salt mixture and owing to the cooling, remained for the most part undissolved during the melting process, and protected the floor of the vessel. On this layer of sodium chloride the finely powdered mixture of NaCl + KCl was placed. The advantages of a composite cylinder of this kind over a crucible made in one piece are obvious, when one considers the importance of durability with regard to temperature differences that occur, and renewals with reference to breakages. For leading in the current an iron fusing-electrode was used, which was connected to the negative pole of the circuit by the end of a cable. The fusing-electrode was then brought so far up to the

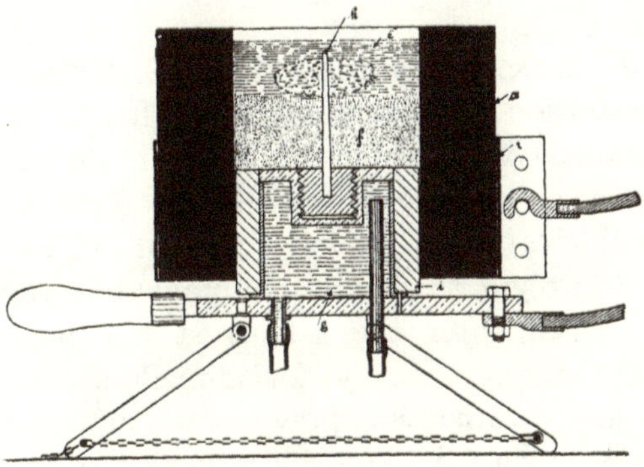

FIG. 49.

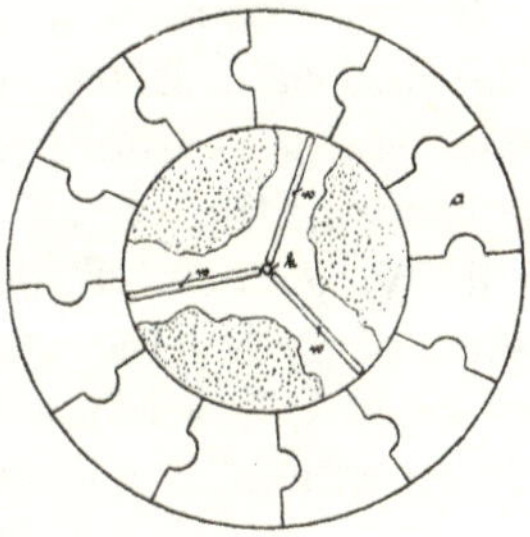

FIG. 50.

anode in the salt mixture with which the vessel was filled that an arc was formed for a short time, and in this manner the salt was quickly melted. By the gradual separation of the upper fusing-electrode from the anode towards the cathode a larger pool of fused salt was soon formed, until it reached the cathode. The fusing-electrode was then removed, and the whole of the salt gradually dissolved. Instead of fusing the salt in this manner by arc heating, it can be reduced to a state of fusion by resistance heating by the insertion of thin carbon rods between the anode and cathode. As soon as a sufficient quantity of fused salt exists, electrolysis can then start after the carbon rods have been removed, simultaneously, if possible. The previously warmed diaphragm was carefully suspended within the clay ring. As the collector vessel a sheet-iron box was used,

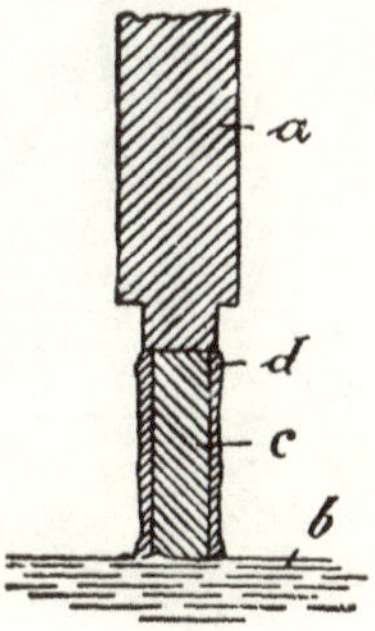

FIG. 51.

from which the sodium was scooped out with a ladle. The chlorine gas escapes through an aperture provided for the purpose in one of the carbon segments.

In an apparatus with which we first succeeded in manufacturing calcium in larger quantities, the melting vessel also served at the same time for the anode, and was constructed in a manner similar to the furnace depicted in Fig. 48. The cathode consists of an iron bar placed in a cooling vessel. The cathode-holder by which the connection to the electrical circuit was made was, of course, insulated from the carbon cylinder by a ring of non-conducting material.

Calcium Furnace of the Elektrochemische Werke Bitterfeld.—For the continuous separation of calcium from the fused salt the Bitterfeld Electro-chemical works raise the cathode at the same rate as the metals are deposited. At the moment of separation the metal solidifies and becomes covered with a solidifying crust of the electrolyte, so that it is obtained in the form of rods (Fig. 51), which, after the salt crust has been knocked off, are for most purposes ready for immediate use. This method of working at the same time prevents the injurious increase of the surface of the cathode, and the reduction of the current density at this part; it was much more easily carried out in this case, as our first experiments already demonstrated the necessity for the separation of the calcium at a temperature as nearly as possible at its melting-point, so as to guard against the calcium re-dissolving as chloride.

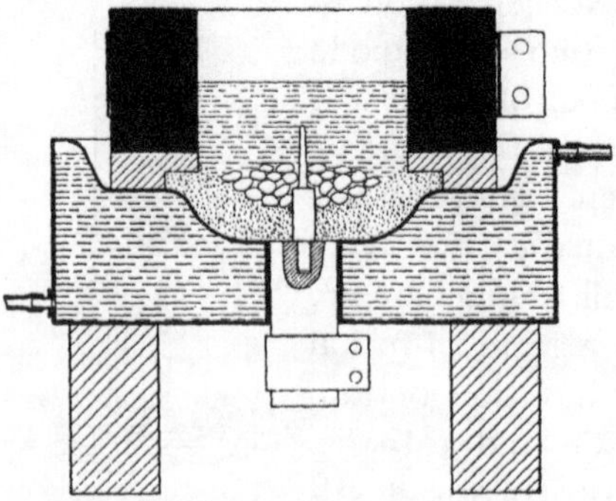

FIG. 52.

Borchers' Strontium Furnace.—For the extraction of strontium the carbon cylinder was retained, but the cooling vessel beneath the floor of the crucible was widened, so that the metal separating out here, only slightly influenced by the stream of chlorine gas at the anode, could sink more rapidly to the bottom and settle on the cooled strontium-chloride floor of the crucible (Fig. 52).

Vincent's Furnace.—For heating *charges which themselves, or whose converted products, remain solid even in the higher temperatures of the electric furnace* (*graphite*), the designs of Vincent, Keneval, Roberts, and Maxim should be considered, although the furnaces were proposed for other purposes.

Vincent[1] published the designs for passing the furnace charge through the circuit traversed by the current; in the one

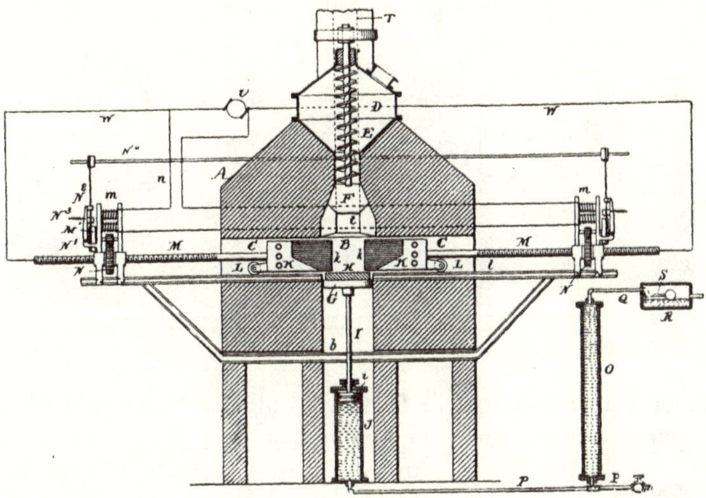

FIG. 53.

the charge was taken through vertically, and in the other horizontally. In both cases the charge is pushed through between the electrodes by means of a worm conveyor. The electrodes are arranged horizontally in the first case and vertically in the second, and the charge then falls into a collecting chamber, or it is gradually carried out by means of a sloping hearth in the furnace (Figs. 53, 54, and 55).

Keneval's Furnace.—In Keneval's Furnace[2] the contacts for the charge, which acts as the electrical resistance, are made in the shape of rollers, the object of which is to provide the motion both for the charge and the products converted from it. The

[1] United States Patent, No. 551,014.
[2] United States Patent, No. 588,866; [English Patent, No. 19,512 (1897)].

idea, good in itself, is, from a constructional standpoint, unsuitably carried out in the particular patent specifications.

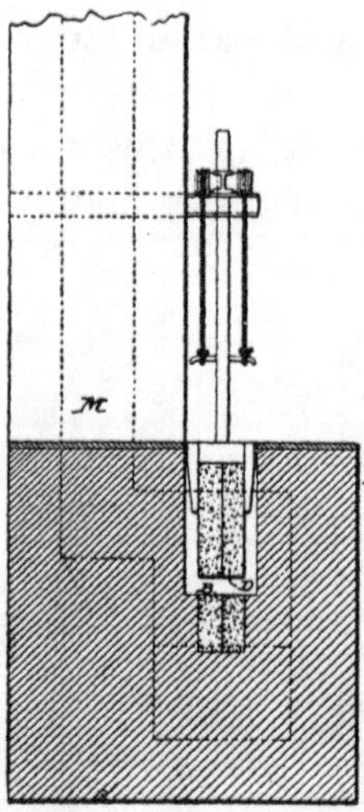

Fig. 54.

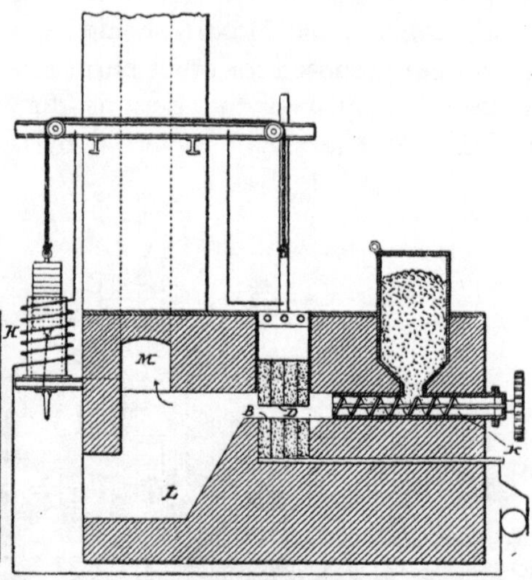

Fig. 55.

Further, such a rolling-mill furnace would hardly be suitable for the purpose for which the inventor intended it, namely, the manufacture of carbide, but would be better adapted for producing substances which remain solid, such as graphite.

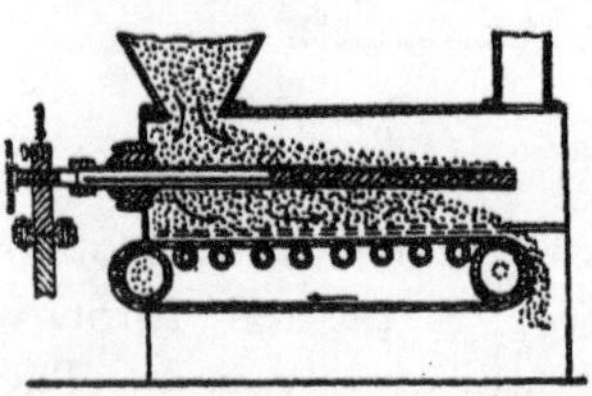

Fig. 56.

The Roberts Furnace.—Roberts[1] conveys the charge by means of a belt conveyor between the electrodes, which are rotatably mounted in one of the furnace walls. They can be brought together to start the heating, and later be revolved apart to the desired distance. The inventor intended the furnace for the manufacture of *calcium carbide*, for which, however, it did not prove suitable. The carbide fuses at comparatively high temperatures and solidifies very easily, so that it remains hanging to the electrodes during the process of working (Figs. 56 to 60).

Maxim's Furnace.—On account of the same defects the

[1] German Patent, No. 100,476; English Patent, No. 17,609 (1897).

Maxim furnace[1] proved useless for the manufacture of calcium carbide. The electrodes, which at the start are pushed right into the charge, are to be gradually withdrawn from the furnace

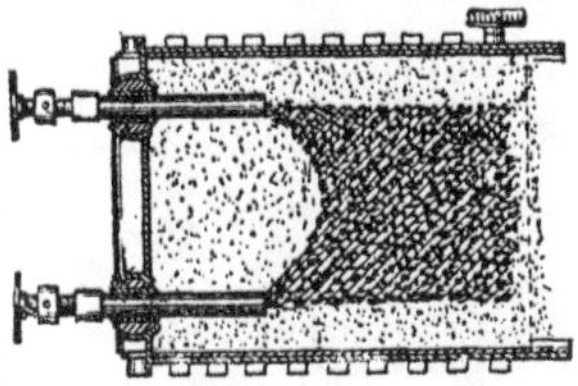

FIG. 57

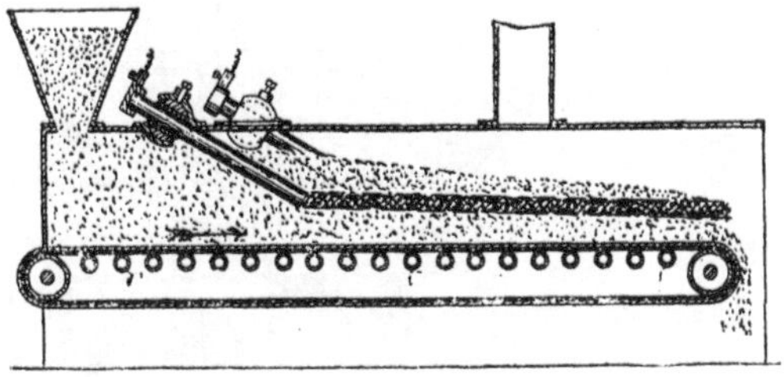

FIG. 59.

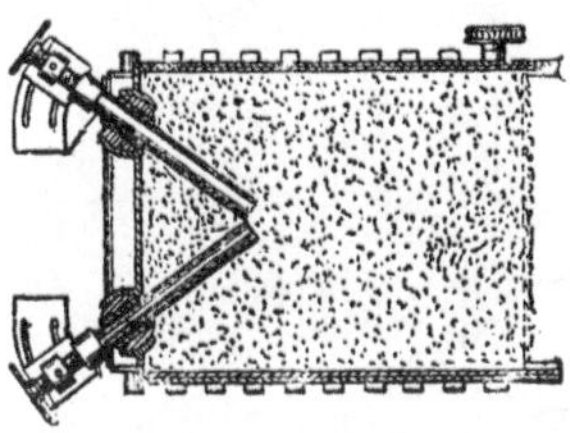

FIG. 58.

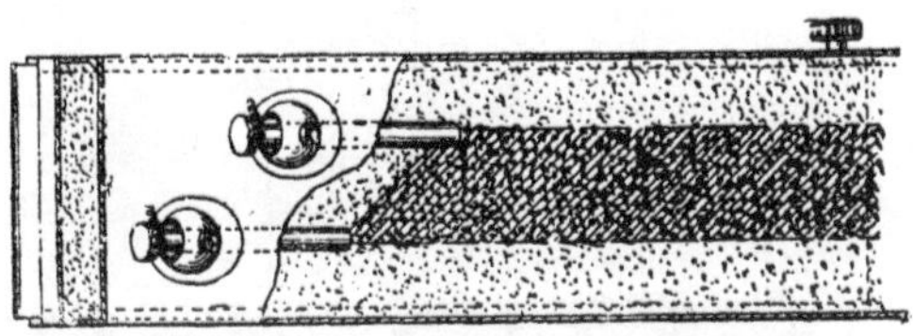

FIG. 60.

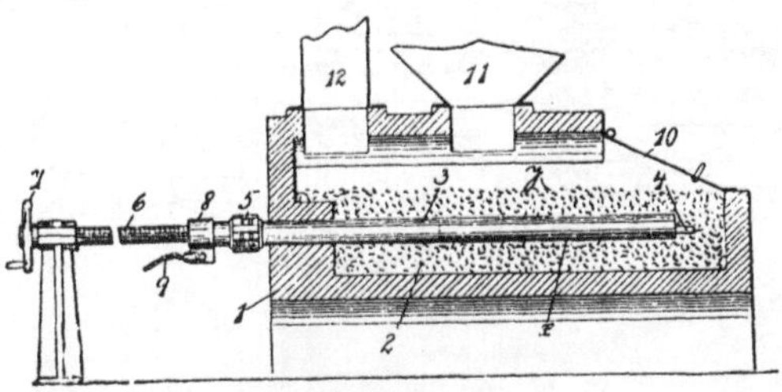

FIG. 61.

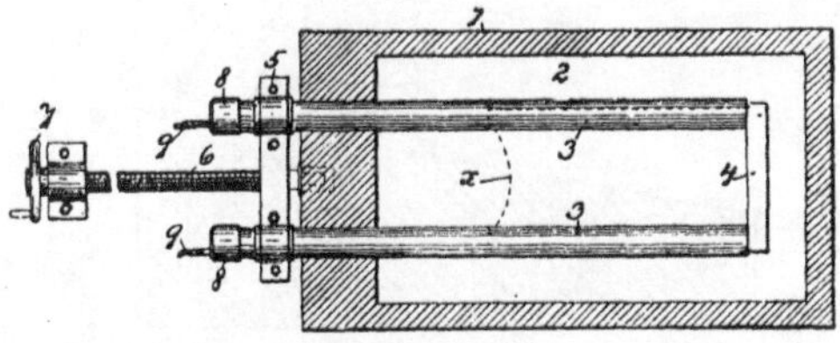

FIG. 62.

at the rate at which the charge is converted, the products of reduction remaining behind (Figs. 61, 62, 63).

[1] English Patent, No. 4075 (1898).

The Hewes Furnaces.—Hewes,[1] in contrast to Maxim, arranges one of the electrodes as a fixture in the furnace, and only

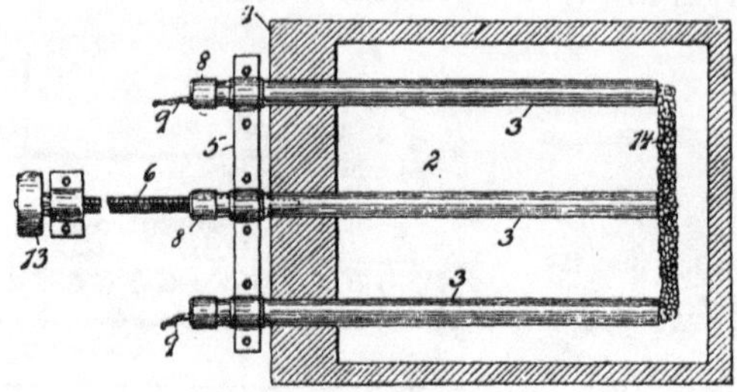

FIG. 63.

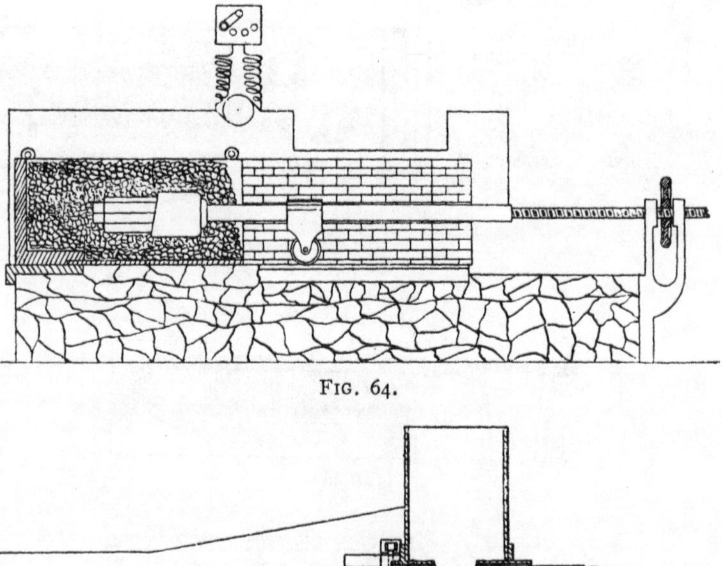

FIG. 64.

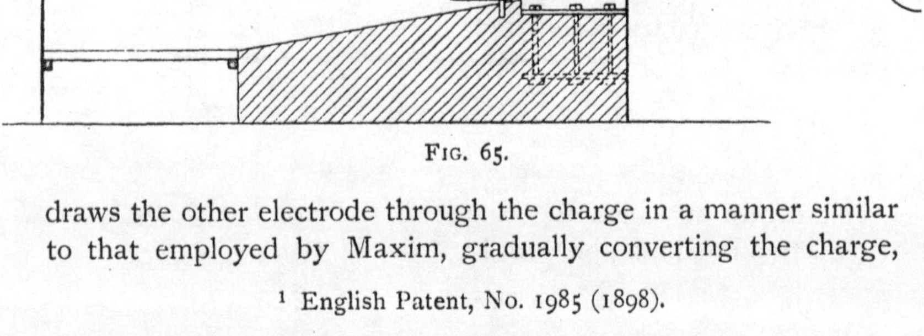

FIG. 65.

draws the other electrode through the charge in a manner similar to that employed by Maxim, gradually converting the charge,

[1] English Patent, No. 1985 (1898).

i.e. its core (Fig. 64). In another design[1] by means of a reciprocating piston he pushes the charge along horizontally between the electrodes, which are mounted on pivots so that they can swing

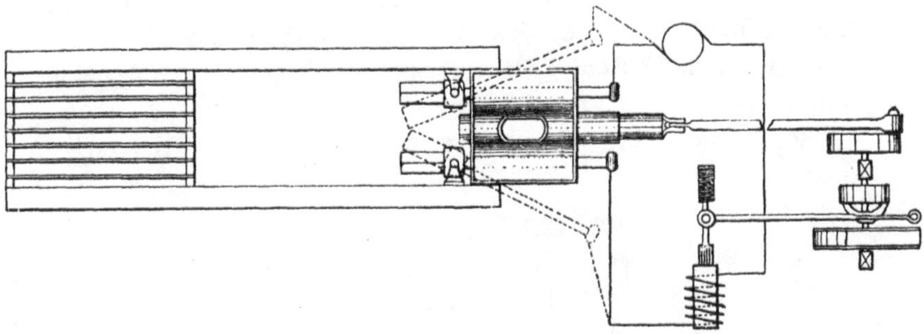

FIG. 66.

apart in a horizontal plane just as far as is required by the charge forced between them (Figs. 65 and 66).

Furnace of Koller and Schuckert & Co. — Koller and Schuckert & Co.[2] propose to improve the method of conveying the current through large electric furnaces with the possible suppression of the current losses in the electrode connections by embedding carbon blocks in the charge as intermediate

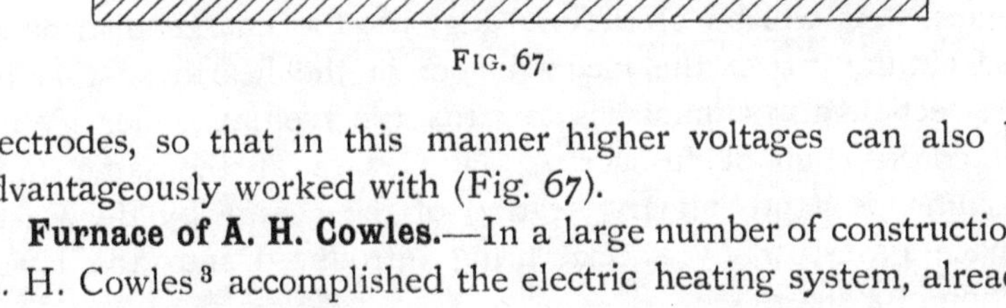

FIG. 67.

electrodes, so that in this manner higher voltages can also be advantageously worked with (Fig. 67).

Furnace of A. H. Cowles.—In a large number of constructions A. H. Cowles[3] accomplished the electric heating system, already

[1] United States Patent, No. 598,318 of February 1, 1898; [English Patent, No. 1983 of January 25, 1898].

[2] German Patent, Nos. 119,463-4-5; English Patent, No. 8894 (1900).

[3] United States Patent, Nos. 750,094-5-6, 750,170-1 of January 19, 1904; German Patent, No. 115.742 of 1898; English Patent, No. 6061 (1899), and

proposed by Borchers in 1898, to arrange the charge and heating resistances so that their cross-sections diminish towards the melting zone, and so that the greatest current density and, therefore, the highest temperature are produced here. Of the large number of furnace constructions described and illustrated in the American patent specifications, I will give the following example:—

The furnace (Fig. 68) has a carbon resistance the cross-

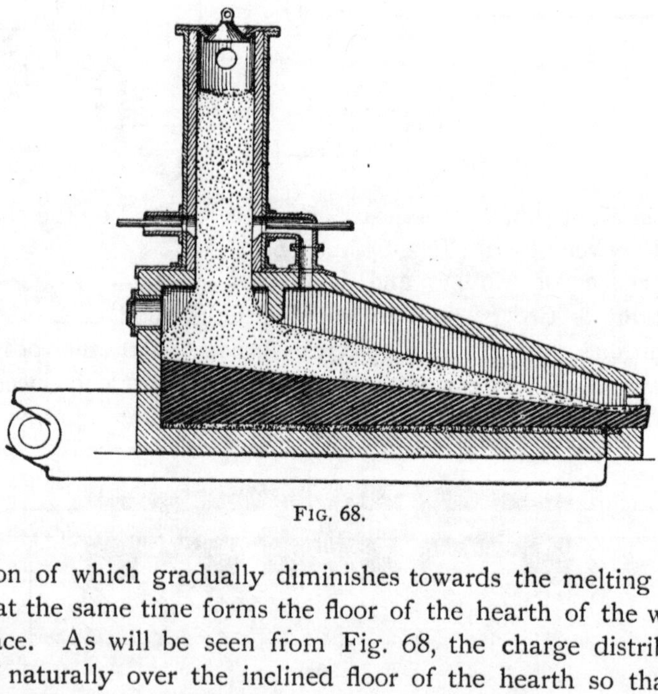

FIG. 68.

section of which gradually diminishes towards the melting zone and at the same time forms the floor of the hearth of the whole furnace. As will be seen from Fig. 68, the charge distributes itself naturally over the inclined floor of the hearth so that its cross-section also diminishes towards the melting zone. With this construction of furnace A. H. Cowles at the same time recommends a preliminary heating of the charge by the waste furnace gases, or by special firing introduced into the lower portion of a shaft.

In the furnace shown in Fig. 69 the diminution in cross-section is attained by the use of a conical pit with the apex

United States Patent, No. 660,064 of 1900; cp. Chapter V., "Combined Resistance and Direct Arc Heating."

downwards, the hearth of which is formed by one of the carbon contacts, while the conductor rods facing each other are inserted in the upper part of the pit. Here also provision is made for a preliminary heating of the charge.

In the same way the cross-sectional reduction is obtained with the construction in Fig.

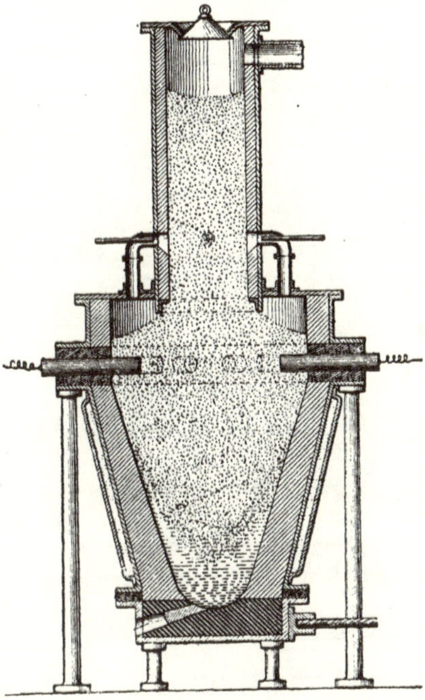

FIG. 69.

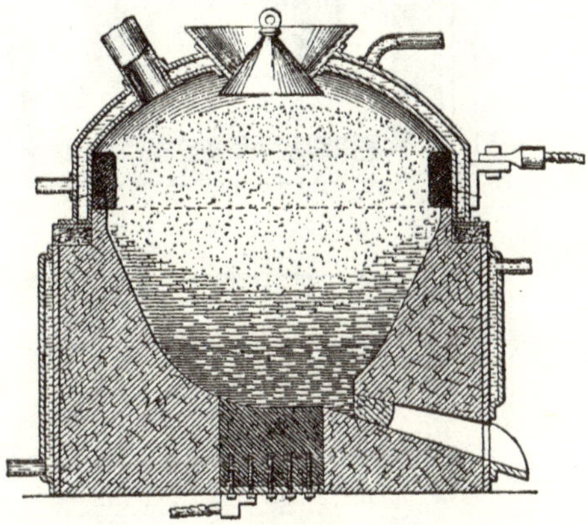

FIG. 70.

70. The top connection to the source of current is established through the water-jacketed dome and a carbon ring embedded in it.

INDUCTION FURNACES.

The Ferranti Induction Furnace.—A new class of electric furnace, embodying ***the most direct method of resistance heating***, was first given by Ferranti.[1] He used closed ring magnets built up of iron plates BF, of which the one limb was wound with wire connected to an alternating current circuit; and at right angles to the lines of force he arranged an oval annular trough of non-conducting material, in which he could melt metals or bring them

[1] English Patent, No. 700 (1897).

to a red heat. By using a metal trough, water and other liquids could be heated in it (Figs. 71, 72, 73, 74).

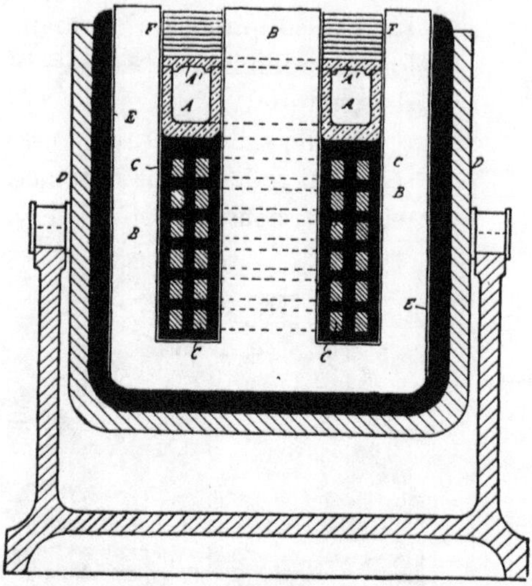

FIG. 71.

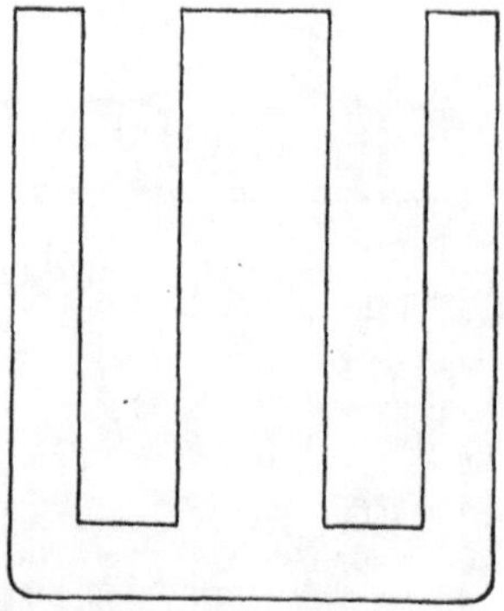

FIG. 72.—Transformer sheet iron.

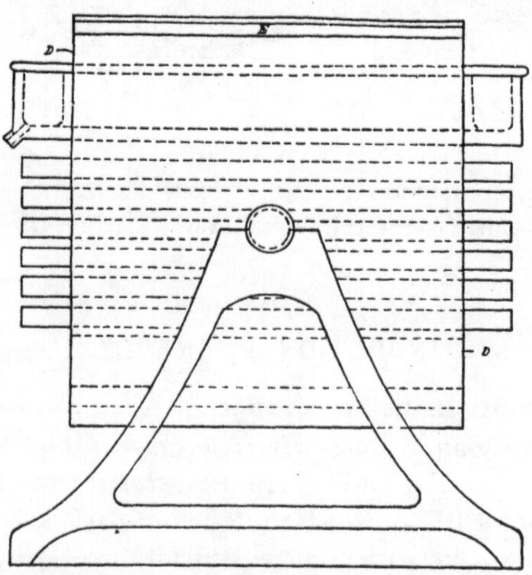

FIG. 73.

With Ferranti's invention, the principles underlying the construction and working of *induction furnaces without electrodes* were so far recognized that rapid progress was now made in their further development ; indeed, up till then all attempts to refine iron electrically had failed, principally on account of the impossibility of controlling the quantity of carbon which the

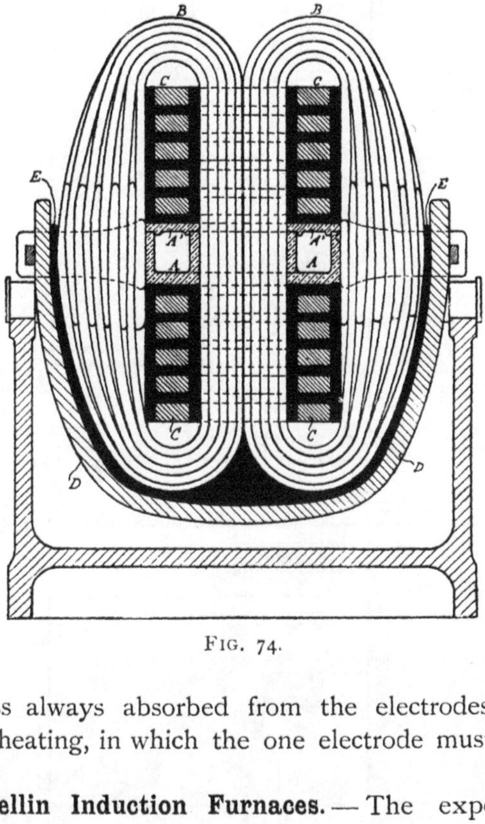

FIG. 74.

molten mass always absorbed from the electrodes, even with electric arc heating, in which the one electrode must at least be of carbon.

The Kjellin Induction Furnaces. — The experiments of F. Kjellin,[1] according to his own communications, led to the following conditions for the construction and working of these furnaces.

The furnace chamber is made in the form of a circular trough, the floor and sides of which consist of masonry. The trough is covered over at the top. In a central pit stands a square core, built up of thin soft iron plates and surrounded by a coil of

[1] English Patent, No. 18,921 (1900), and United States Patent, No. 682,088 of September 3, 1901 ; German Patent, No. 126,606 of October 2, 1901.

insulated copper wire. The core projects above the furnace chamber and forms a rectangle which, in relation to the furnace chamber, occupies the same position as one link of a chain to

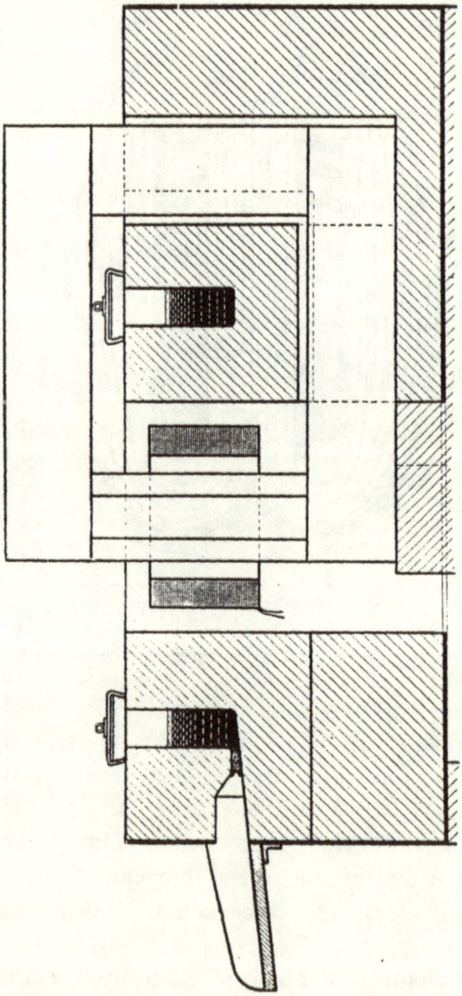

FIG. 75.—Scale $\frac{1}{30}$.

another. The wire coil is connected to the terminals of an alternating current generator.

During the passage of the alternating current through the coil a magnetic field is created in the iron core. This magnetic field is continually varying both in intensity and direction, and by its action on the metal contained in the furnace chamber it produces an alternating current in the annular metal bath. The bath has only one layer right round the core, and therefore *the strength of the current in it is very nearly equal to the current from the generator multiplied by the number of wire turns in the coil.* The voltage naturally diminishes with increase of current.

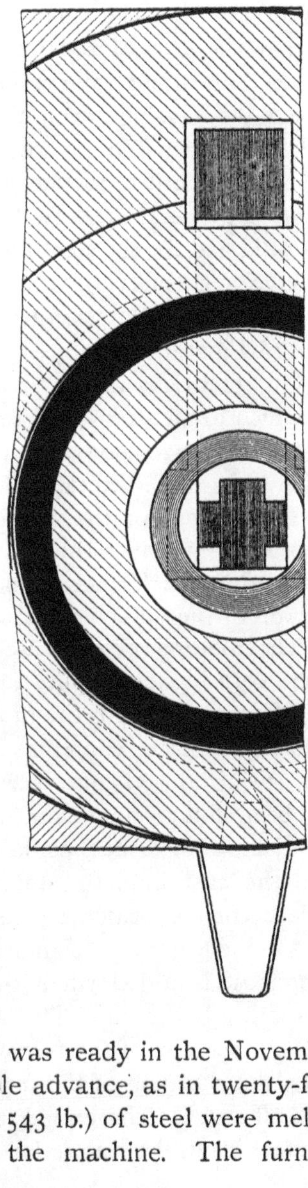

FIG. 76.

In this manner a generator for high-pressure alternating current can be used, and, without having to employ energy-absorbing electrodes and heavy copper conductors, a low-voltage alternating current of high current intensity can be obtained in the furnace.

Towards the end of February of 1900 the first steel furnace was ready in Gysinge, Sweden, so that experiments could be made with it. The time taken by the experiments was not long, for already, on March 18, the first casting was obtained, and the steel proved to be of excellent quality.

The question was, therefore, technically but not economically solved, as with the 78-k.w. dynamo used not more than 270 kg. (595 lb.) of cast steel were obtained in twenty-four hours, and the furnace capacity was only 80 kg. (176 lb.).

It was, therefore, necessary to construct a new furnace having a larger iron core, and this one, which was ready in the November of 1900, proved to be a considerable advance, as in twenty-four hours from 600 to 700 kg. (1323–1543 lb.) of steel were melted with an output of 58 k.w. from the machine. The furnace capacity was 180 kg. (397 lb.), and the charges of 100 kg. (221 lb.) took from three to four hours.

The yield per H.P. was, however, far from satisfactory in consequence of the magnitude of the cooling surfaces of the walls in proportion to the quantity of heat developed in the furnace, so that the cost of repairs, reckoned per ton, was considerably higher than would have resulted with a larger furnace.

As the Gysinge Sulphite Works were burnt down on the 11th of August, 1901, it was decided to erect steel works in their place, and to utilize for the same the water-power which had been acquired there. A 300-H.P. turbine direct coupled to a generator was obtained for the steel furnace. The new furnace is calculated to have a capacity of 1800 kg. (3970 lb.), and the output is to amount to at least 1500 (metric) tons per annum with charges of cold, raw material.

Quartz bricks are to be used at first for the walls, as for steel for which great strength is required acid walling is most suitable; experiments are, however, to be made with magnesite bricks, which have the advantage of being more fire-proof.

The voltage of the alternating current generator is fixed for 3000 volts, to reduce the amount of copper. The same pressure was used in the former furnaces, and on account of the protected position of the inducing coil, did not prove injurious.

The melting process used at Gysinge, where the production of the very highest quality tool-steel is aimed at, is as follows:—

After running off the molten metal, when a little more than half the furnace contents is removed, a charge of cast iron is first introduced, and then as much scrap iron is added as has been found necessary from the experience gained in the manufacture of a steel containing the desired quantity of carbon from the cast iron, the amount of carbon of which has meanwhile become considerably smaller.

After everything is molten and is fairly super-heated, ferromanganese is added, when the super-heating must be continued for another half-hour. The steel is then ready to be tapped.

The top edge of the furnace is on a level with a platform from which the charging of the furnace is conducted; the cover is first lifted off and the charge fed in. The heat being generated within the steel itself, the slag is far less hot than in a Martin furnace, so that no particular inconvenience is caused by the heat during the operation of charging.

Repeated trials with a furnace taking 225 E.H.P. have yielded outputs of 4100 kg. (4·035 tons) in 24 hours. With his 300-H.P. furnace, now probably already at work, Kjellin hopes to attain an output of 20 kg. (44·1 lb.) of steel per electrical H.P. in 24 hours.

The first furnace (225 E.H.P. = 168 k.w.) is represented in Figs. 75 and 76 on a scale $\frac{1}{30}$ full size.

Frick's Induction Furnace.—In the endeavour to bring the winding of the primary circuit as near as possible to the secondary circuit, according to communications by Frick, the inner wall of the ring hearth was made as thin as possible, whereby, however, the durability of this furnace wall was considerably reduced. These and other defects (too high self-induction and too large phase differences) can, according to Frick,[1] be removed by placing the winding of the primary circuit above the secondary, *i.e.* above the smelting hearth (Fig. 77).

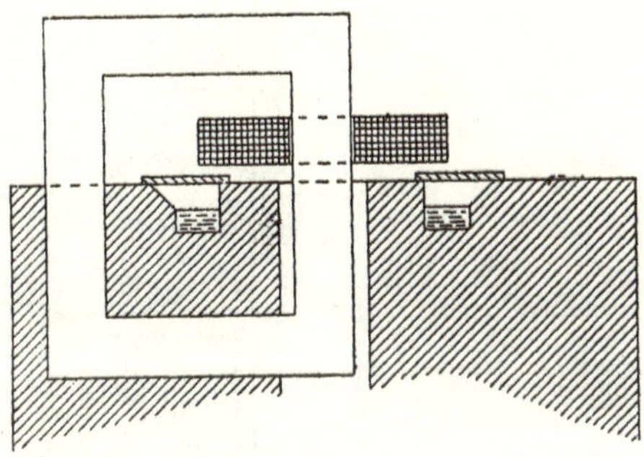

FIG. 77.

Hjorth's Induction Furnace.—A. Hjorth, as appears from a direct communication of his on his double-hearth furnace, seems to place less value on the great closeness of the winding of the primary circuit to the secondary circuit. In fact, he arranges it outside the periphery of the latter, so as to be able to alternately supply two hearths with one transformer, and thus reduce the installation costs and the interruptions to working in smelting works. The electrical part of the plant is the most costly. Should one of the hearths have to be freshly walled, as happens in furnaces in which the molten mass has to be kept at a temperature of about 2000° C., with Hjorth's construction in a short time the transformer can be placed through the next hearth, which, before it is set to work, is filled with molten metal from the other hearth, so that it is ready for immediate use (Fig. 78).

[1] English Patent, No. 4866 (1904).

Schneider's Induction Furnace.—It is unnecessary here to go into the details of the induction furnaces of Schneider, which

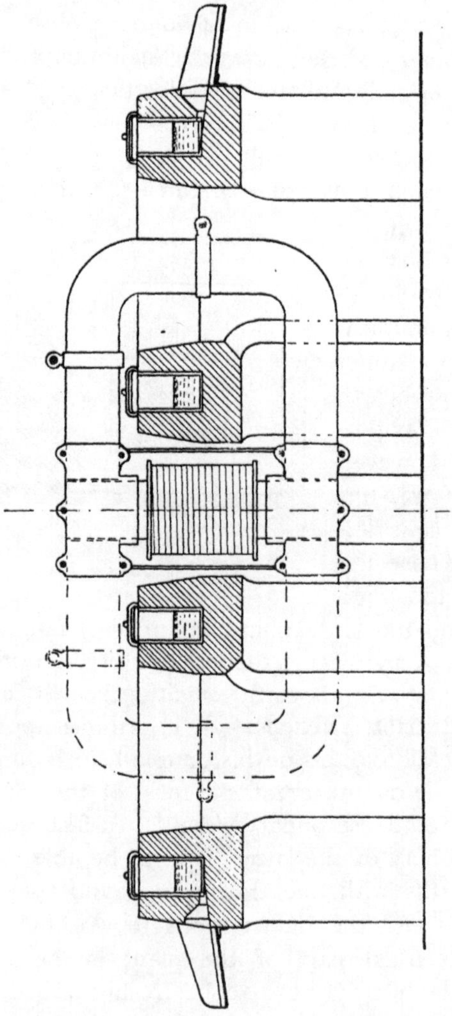

Fig. 78.

are less simple in their whole construction and are doubtful in operation, as they do not possess any advantages in contrast to the simple designs of Kjellin, Frick, and Hjorth.[1]

[1] United States Patent, No. 761,920 of June 7, 1904, and No. 763,330 of June 21, 1904.

CHAPTER III

INDIRECT RESISTANCE HEATING

FURNACES IN WHICH THE SUBSTANCE TO BE HEATED IS IN CONTACT WITH AN ELECTRICALLY HEATED SUBSTANCE

Introduction.—If we consider the furnace represented in Fig. 3 from the point of view that the main object was to test the behaviour of the diamond powder enclosed in the crevice of the wire during the heating, we can claim it as the prototype for this method of heating, for the iron wire was there heated direct by the current.

As the second attempt in this direction we should then record the small furnace of Despretz in the year 1849; it is illustrated in Fig. 79 to a scale twice full size. According to his account in the *Comptes Rendus*, vol. xxix., of the year 1849, the heating resistance consisted of a tube of sugar-carbon, 7 mm. wide and 23 mm. long, which was closed by means of two plugs of the same material and connected to a heavy current circuit. The substance to be heated was contained inside the tube.

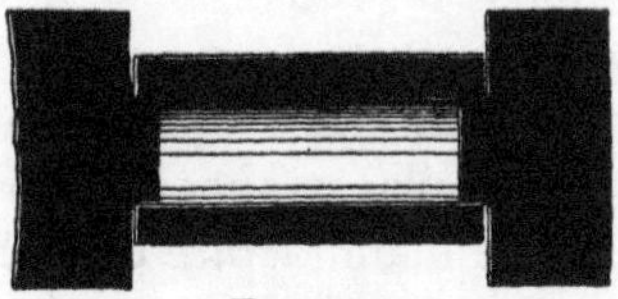

FIG. 79.

The last design of the Cowles furnace, illustrated in Fig. 7, at the commencement of the operation, would also belong to this second furnace group, as during the first moments it is the pair of carbons projecting beyond the carbon bundle that is heated as the resistance material.

Borchers' Furnace for Indirect Heating.—Finally, in the year 1891 I published a description and sketch of an experimental furnace which works exclusively on the principle of this second method of heating. The construction, now universally known from various publications, is on the following lines: Inside a heating chamber of fireproof stone a thin carbon rod bridges between the massive carbon poles of an electrical circuit (Fig. 80). The furnace charge is packed round this thin carbon

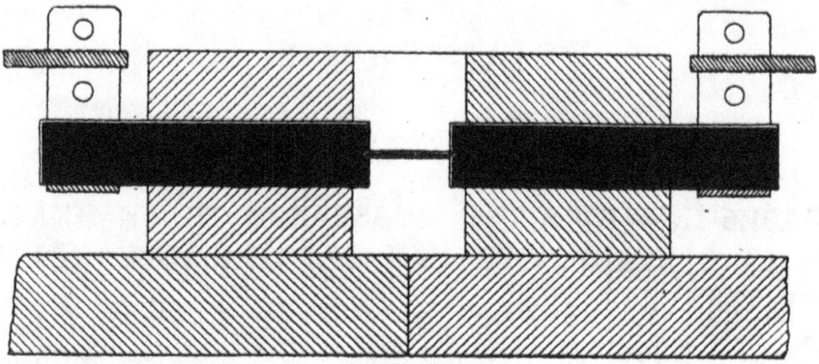

FIG. 80.

rod, which becomes strongly heated and heats the charge from within. Already in my first publication on this subject, in the first edition of my book "Elektrometallurgie" (1891), I pointed out that with this method of heating *all oxides hitherto regarded as irreducible could be reduced by the aid of carbon heated in this manner.* I then specially mentioned the reduction of silica as well as the fact that an absorption of carbon by the products of reduction (as far as they combine or alloy with carbon) was not to be avoided (consequently, the uselessness of this apparatus for the production of pure aluminium and other metals which behave in a similar manner towards carbon, but the utility of the same for the manufacture of carbide).

Acheson's Carborundum Furnace.—That this method of arranging the resistance material can be easily applied on an industrial scale we already see from the patent specifications of Acheson[1] for the production of carborundum in the year 1892. The contacts consist of bundles of carbon rods, B, which are held in metal frames B_2, B_3 by means of strips of copper, and they are connected to the main circuit R (Figs. 81, 82).

A core C, comprised of powdered coke, is stamped down into

[1] German Patent, No. 76,629, October 16, 1892; [English Patent, No. 17,911, October 7, 1892].

the charge and forms the heating resistance; in the vicinity of the poles it widens considerably so as to protect the rods forming the conductor poles. The foundations A and the front and back walls of the furnace which support the poles are built up with

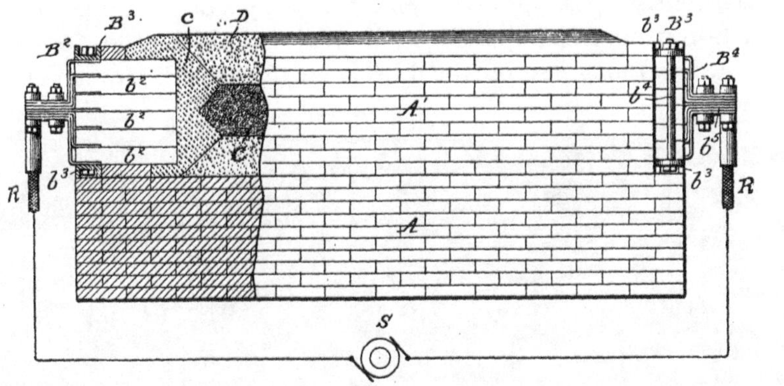

FIG. 81.

mortar, whereas the side walls A′ of the furnace are constructed of stonework placed together dry (uncemented) each time the

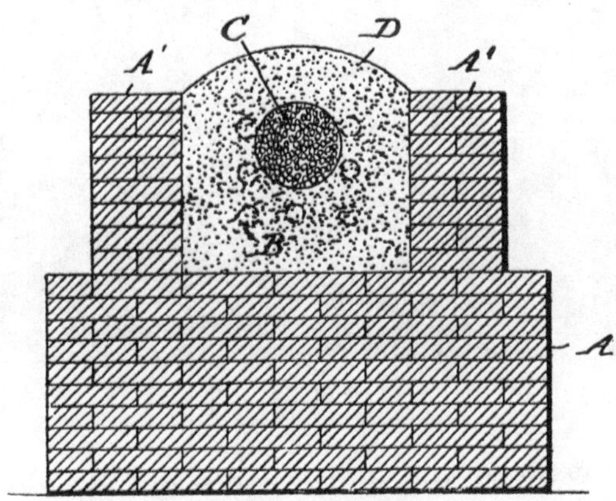

FIG. 82.

furnace is freshly charged, exactly as the whole of my experimental furnace is built up with a few stones.

Figs. 83–86 give views of the present-day furnaces in different stages of working. According to Fitzgerald, the total length of a furnace is 7000 mm. (22 ft. 11½ in.), the inside length is

5000 mm. (16 ft. 5 in.), the breadth 1800 mm. (5 ft. 11 in.), and the height 1700 mm. (5 ft. 7 in.). Twenty-five carbon rods, each 860 mm. (2 ft. 9$\frac{7}{8}$ in.) in length and having a section of 100 mm. × 100 mm. (15$\frac{1}{2}$ sq. in.), are arranged in five horizontal rows, each consisting of five bars placed side by side. Copper plates, each 100 mm. (4 in.) wide, 13 mm. ($\frac{1}{2}$ in.) thick, and 660 mm. (2 ft. 2 in.) long, are provided between them at the front. They

FIG. 83.

are clamped between wide iron plates, and to the outwardly projecting copper plates are connected the circuit cables by means of bolts. These contacts lie in solid masonry walls, whereas the side walls are built up of masonry loosely put together without mortar to approximately the height of the carbons when the furnace is to be started. After the charge is put in, iron plates are inserted to keep the charge away from the carbon conductors, so that the empty spaces can afterwards be filled in with small coke. The charge is hollowed out in the shape of a semi-circular trough into which rather coarse lumps of small coke free from dust are stamped, and the whole is rounded off at the top to form a cylindrical core about 530 mm. (1 ft. 9 in.) in diameter. The empty spaces round the carbon poles are then filled in with small coke, and the rest with charging material.

The current for heating is supplied at pressures varying from

FIG. 84.

75 to 210 volts, the total power taken amounting to about 746 kilowatts. It is, of course, clear that at the start of the operation only a small current at a high voltage will be required, whereas later the current will be large and the voltage will drop.

The charge, which is always weighed out and mixed in loads of 500 kg. (1102 lb.), contains in this weight 261 kg. (575 lb.) of sand, 177 kg. (390 lb.) of coke, 53 kg. (117 lb.) of sawdust, and 9 kg. (20 lb.) of common salt.

As already pointed out, each furnace takes 746 k.w., and with this consumption of power yields 3150 kg. (3·1 tons) of carborundum in thirty-six hours, corresponding to 8·5 k.w.-hours per kg. of carborundum (3·86 k.w.-hours per lb.). As the Carborundum Works at Niagara Falls have three generator units of 746 k.w. each, and for each unit there are five furnaces installed, 9450 kg. (9·3 tons) of carborundum can be turned out

FIG. 85.

in thirty-six hours, giving a daily output of 6300 kg. (6·2 tons) of carborundum.[1]

Acheson's Graphite Furnace.—At the present day Acheson also utilizes these furnaces for the production of graphite from amorphous varieties of carbon. In the manufacture of carborundum he already noticed that the coke in the heating core became partially converted into graphite, and after making numerous experiments be believed that he had found a method of expediting the formation of graphite in the electric furnace by means of "metal vapours." In his first patent specification[2] referring to this, he mentions iron, silicon, titanium, and borium as specially effective.

In a publication of observations carried out by me in the

[1] For further particulars about carborundum, see "Carborundum," by Francis A. J. Fitz-Gerald, Chemist of the International Graphite Co., Niagara Falls, New York.

[2] United States Patent, No. 711,031, October 14, 1902.

years 1887–1890, in addition to the substances just enumerated, I also alluded to aluminium, the cerite metals, manganese, and many other substances, briefly, all those substances which, on reduction from their oxides through the agency of electrically-

FIG. 86.

heated carbon, have the tendency to form carbides, which again become decomposed with temperature or pressure variations, the metal becoming volatilized, and the carbide carbon remaining as graphite. I, therefore, concluded that if I impregnated the

FIG. 87.

carbonaceous mass to be heated with small quantities (fractional percentages of the carbon to be worked with) of reducible compounds of such substances, in part enumerated above, which form carbides with carbon, alloy with carbon, or the carbides of which can alloy with carbon, these substances, similar to the substances designated in many chemical processes as contact or catalytic substances, probably on account of the alternative formation and decomposition of carbides and compounds of these with carbon in conjunction with refining processes, gradually bring the carbon to the crystalline, graphitic state natural at the exceedingly high temperatures at which the reduction of the oxide and the dissociation of the carbide mostly take place.

This was meanwhile confirmed by the experiences of Acheson,[1]

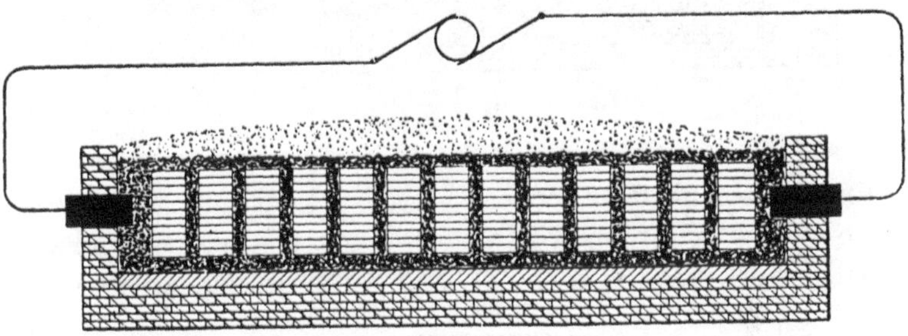

FIG. 88.

according to whom carbon containing particles of ash, or intentionally added ingredients of silicic acid, clay, alumina, magnesia, chalk, or peroxide of iron, can be easily converted into graphite by electric heating (magnesia excepted; on the other hand manganic oxide). Later on experiments in my laboratory[2] proved that aluminium oxide is the most effective, then manganic oxide and peroxide of iron.

For the production of graphite a mixture of carbon, containing the graphitizing ingredients, is packed round the carbon core. The furnace is made longer (9000 mm., 28 ft. $8\frac{1}{4}$ in.), but is somewhat narrower (500 × 350 mm., 1 ft. $7\frac{3}{4}$ in. × 1 ft. $1\frac{3}{4}$ in.)

[1] United States Patent, No. 568,323 of September 29, 1896.

[2] Borchers, "Institut für Metallhüttenwesen und Elektrometallurgie," 1903; and Weckbecker in "Metallurgie," 1 p. 137 (1904).

than the carborundum furnaces, and it works with about 800 k.w. (Fig. 87).

Following the usual practice, the graphite slabs or rods (electrodes) are compressed with the graphitizing substance only; they are then subjected to a preliminary baking in ordinary furnaces and, packed with granulated coal, are then roasted to completion in electric furnaces (Figs. 88 and 89).

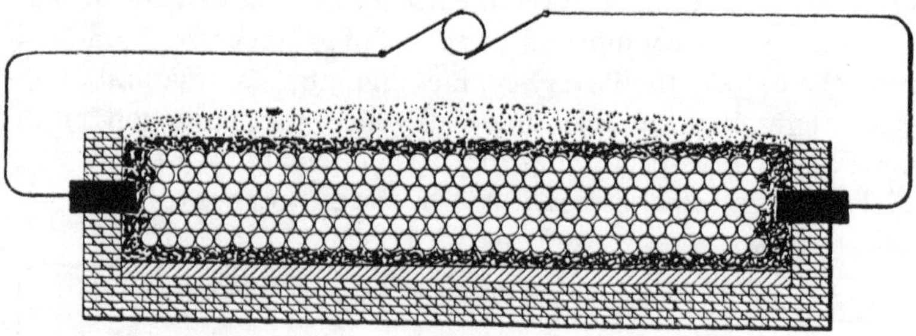

FIG. 89.

Price's Graphite Furnace.—Price adopts the arrangement shown in Figs. 90, 91, and 92. In a furnace, the hearth of which is partially constructed of masonry, and partially built up of courses

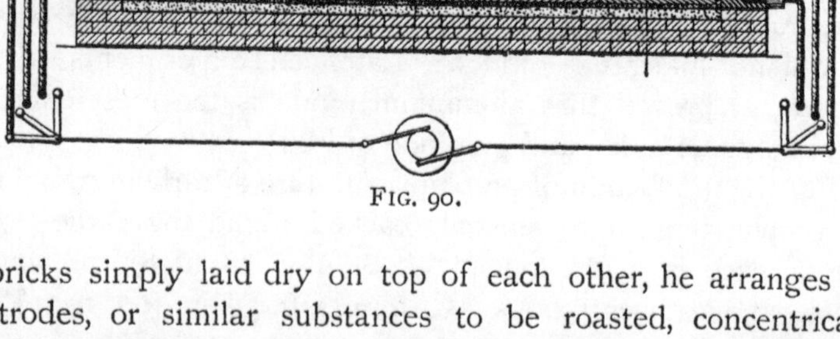

FIG. 90.

of bricks simply laid dry on top of each other, he arranges the electrodes, or similar substances to be roasted, concentrically round a core of coarsely granulated coke, or other suitable electrically-conducting resistance material. Embedded in the end walls of this hearth are carbon blocks which form the contacts, of which the centre blocks are to feed the current to the carbon core, which serves as the first heating resistance, while the carbon rings arranged concentrically round the centre block, but insulated from it by non-conducting strata, are for the purpose of conducting the current to the substances to be heated. These latter are also insulated not only from the carbon cores, but from the neighbouring layers by means of non-conducting strata. The whole mass is embedded within the hearth in refractory, non-conducting material (chalk or sand).[1]

The *modus operandi* is as follows: A current of suitable intensity is first sent through the carbon core, and later, by means of a change-over switch, either the whole current or a greater part

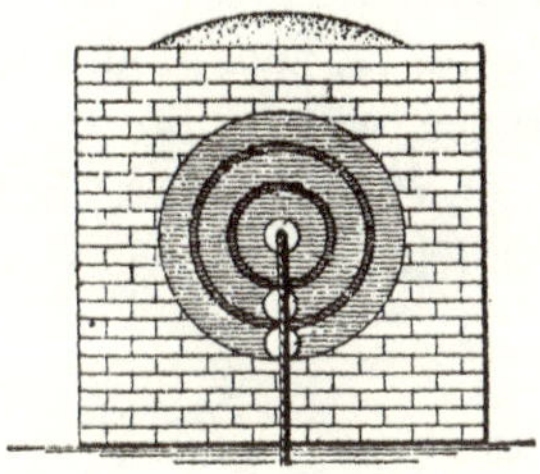

FIG. 91.

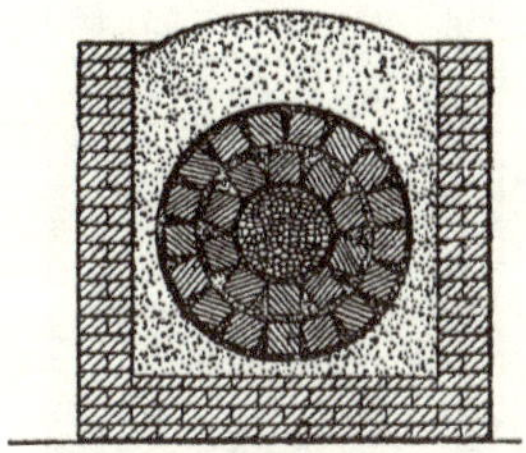

FIG. 92.

of it is conducted through the layers of the substances which are to be raised to a red heat.

In a second patent specification Price describes another arrangement of the substances to be electrically acted upon, and gives a somewhat altered process for heating them. Here also he makes use of a carbon core, but arranges the objects to be heated in the form of a rectangular prism within the furnace, with the carbon core in the centre. Between the separate slabs he packs small anthracite coal so that at first they are insulated from each other and from the carbon core, which acts as the resistance material. The heating is first started through the medium of the central carbon core, when the intermediate layers of small coal become gradually converted into coke and become conductive, so that by degrees further layers of the electrode slabs embedded round the carbon core are being constantly introduced

[1] United States Patents, No. 752,357 and No. 752,358 of February 16, 1904.

into the current circuit as heating resistances, with the result that the heat gets generated in these bodies themselves.

Tone's Silicon Furnace.—Tone[1] arranges resistances for heating larger masses as shown in Figs. 87–90, when *difficultly reducible oxides are to be worked*, the ingredients of which volatilize at temperatures which are near the temperatures of reduction of the oxides and, in addition, readily combine with carbon to form carbides.

A furnace of the simplest kind is represented in Figs. 93 and 94. It consists of a shaft furnace, 2, the bed of which, 3, forms the roof of the collecting chambers, 6, in which the molten mass accumulates and from which it is tapped. The heating resistance, which, like the contacts, 7 and 8, is also embedded in the charge, consists of a

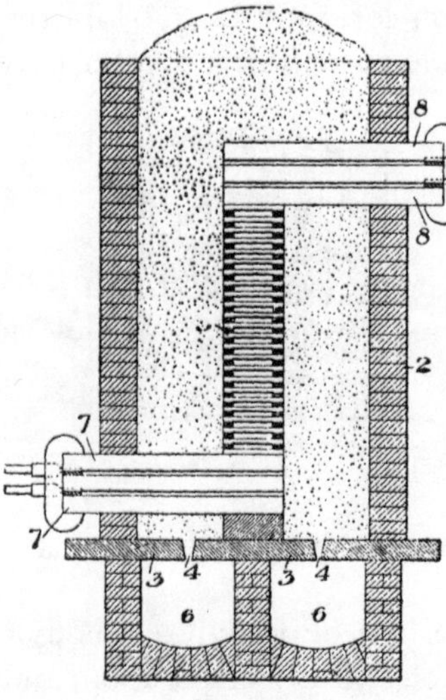

FIG. 93.

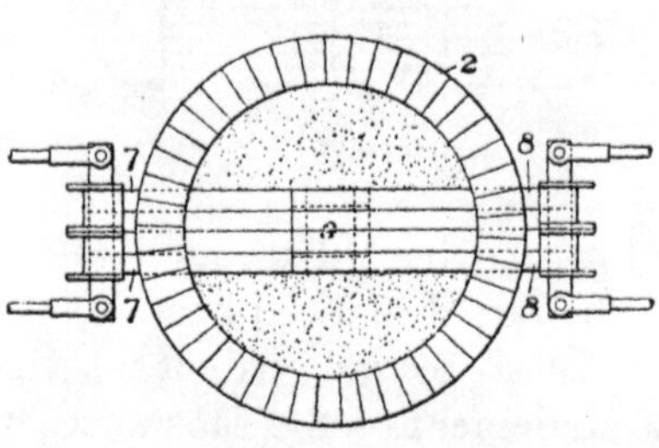

FIG. 94.

column, 9, built up of briquets of carbon. Although this resistance may reach a higher temperature than, for example, the temperature of volatilization of silicon in the reduction of silicic acid, the temperature of the charge can still be kept down sufficiently low to enable the condensing silicon to descend in good time into the colder portions of the furnace and overflow through the openings, 4, into the collecting chambers. The apparatus is also intended for use in the reduction of aluminium oxide. The method to be adopted for the arrangement of the resistance materials in large furnaces is illustrated in Figs. 95 and 96.

[1] English Patent, No. 23,964 (1903).

If, in discussing the last furnaces, I have somewhat departed from the chronological order of publication of resistance furnaces, this has been done with the intention of giving together the con-

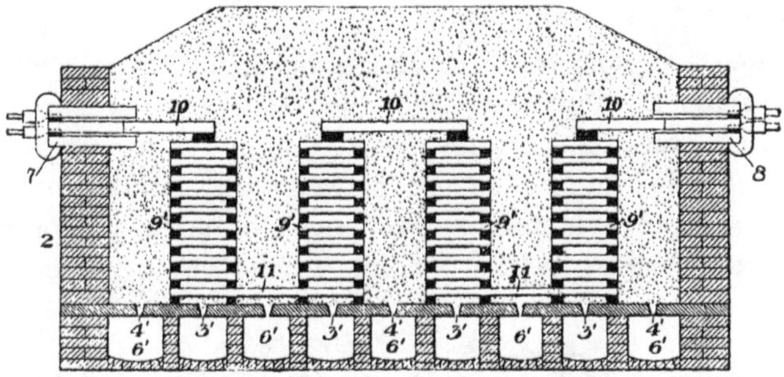

FIG. 95.

structions for the manufacture of carborundum, graphite, and silicon, which have emanated from one source. I will now proceed with those furnace designs which have been passed over until

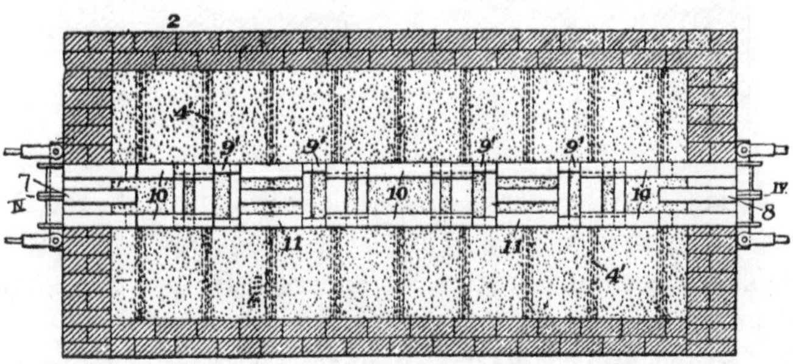

FIG. 96.

now, and will then continue with the so-called closed furnaces, which shall also be treated together.

Elihu Thomson's Furnace.—In an American patent specification[1] Elihu Thomson gives methods for arranging and using resistance materials which are less simple than those already experimentally tried in the simplest form by Pepys in 1815 (split metal wire), by Despretz in 1849 (carbon tube), and by

[1] United States Patent, No. 513,602, January 30, 1894.

Borchers in 1887 (carbon bars), and thereby complicates the preparation of experimental electric furnaces.

The Crompton and Dowsing Furnace.—Calcining furnaces heated by the agency of metal wires and carbon bars embedded in the plates of the furnace floors, as proposed by Crompton and Dowsing,[1] have not proved successful in practice.

Hasslacher's Furnace.—For the conversion of *emery* into *corundum*, Hasslacher starts the operation with fragments of coke embedded in the charge between the carbon conductors as the resistance material. He tries to make the operation continuous by covering an opening in the furnace floor with a

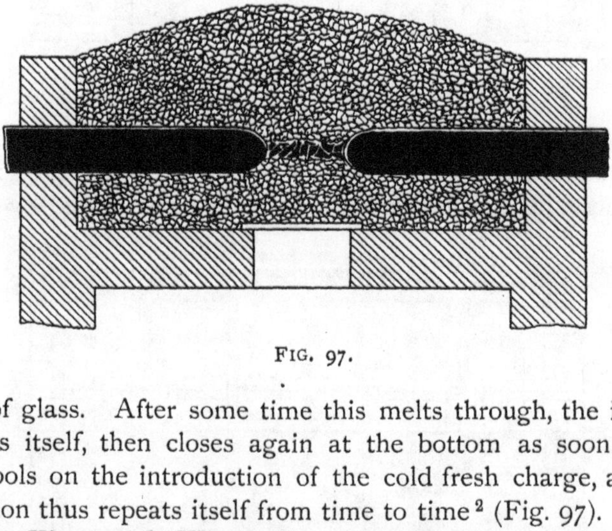

FIG. 97.

plate of glass. After some time this melts through, the furnace empties itself, then closes again at the bottom as soon as the slag cools on the introduction of the cold fresh charge, and the operation thus repeats itself from time to time[2] (Fig. 97).

The King and Wyatt Furnace.—The extremely simple smelting furnace constructed by King and Wyatt[3] consists of a piled-up heap of the charge, and through the charge is passed a thin vertical carbon rod which serves as the heating resistance and bridges the thick carbon leading-in poles (Fig. 98).

Maxim's Furnaces.—In several patent specifications Maxim[4]

[1] English Patent, No. 259 (1893).

[2] German Patent, No. 85,021 of 1895.

[3] United States Patent, No. 562,402 (1896); [English Patent, No. 13,881, June 3, 1896].

[4] English Patent, No. 25,611 (1896); No. 18,989 (1897); German Patent, No. 100,477 (1897).

gives a number of arrangements of connecting the resistance materials to the circuit, and of replacing them as they wear away by feeding them into the furnace. The latter operation is

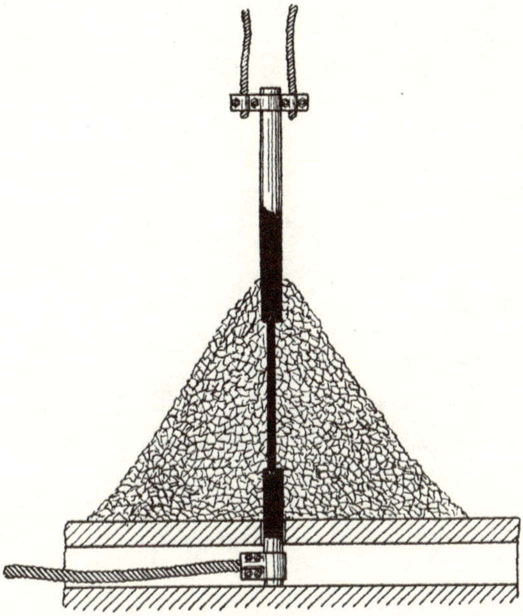

FIG. 98.

carried out through hollow contacts by means of stuffing-boxes, screws, and rams.

Two examples of the arrangements adopted by Maxim may be shortly described. One of the furnaces (Fig. 99) is built in the form of a long rectangular hearth, in which a common contact is arranged in the form of a carbon plate placed vertically along one of the furnace side walls. Through the opposite side wall six thick carbon rods, called electrodes, pass horizontally into the furnace. The outside extremity of each electrode is joined to one side of a current generator by means of a terminal, cable, change-over switch, and an ammeter arranged on a switchboard. Between each electrode and the oppositely situated carbon plate a heating resistance in the form of a thin carbon rod is held by means of a spring under pressure. The spring rests against the switchboard, and is supported on an iron rod rigidly attached to the outer end of the electrode. The electrode is further held by means of the iron rod which passes through the switchboard, and is prevented from turning by a catch which engages in a longitudinal groove in the rod. The latter has a screw-thread cut on it for a part of its length, and is provided

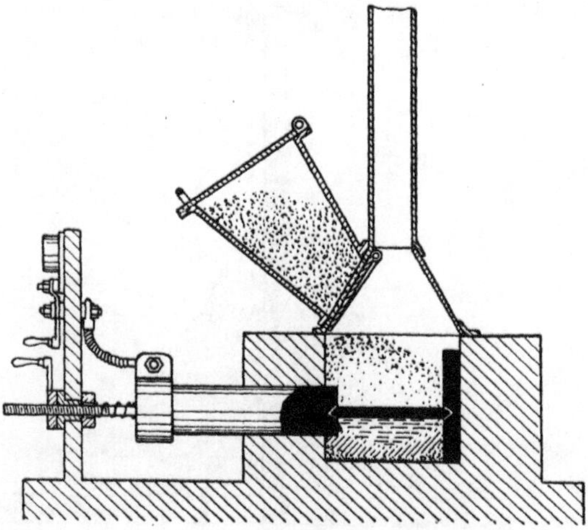

FIG. 99.

with a handle which, when rotated, enables the electrode to be drawn out of the furnace.

A hood with a stack serves to carry off the waste gases, and a feed-hopper is provided for the reception of the material to be handled.

In the second furnace (Fig. 100) the position of the electrodes is regulated by means of a hand-wheel, and reliable contact

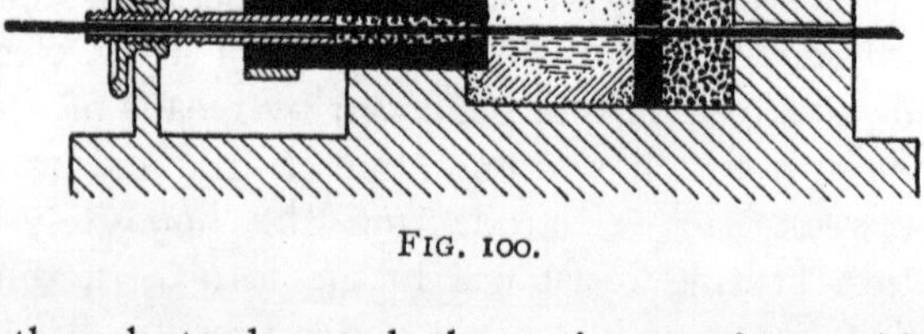

FIG. 100.

between the electrodes and the resistance is provided for by placing powdered coal, or coal-dust, in a chamber at the back of the common electrode plate and enclosing it within the front electrode.

The Patten Furnaces.—In several American patent specifications Patten[1] describes vertical and horizontal arrangements of the heating resistances inside cylindrical furnace pits. In the one the carbons are disposed in a ring round a central rod, the only object of which is to keep the middle portion of the pit free from the charge. In another furnace, in which a central carbon rod functions as the heating resistance, he attempts, by means of a solenoid placed concentrically round the pit, to draw out the heating zone in a manner similar to that employed in arc furnaces. In a third furnace the resistance rods project horizontally from a vertical central rod towards the conducting wall-lining. They are not just arranged one above the other, but

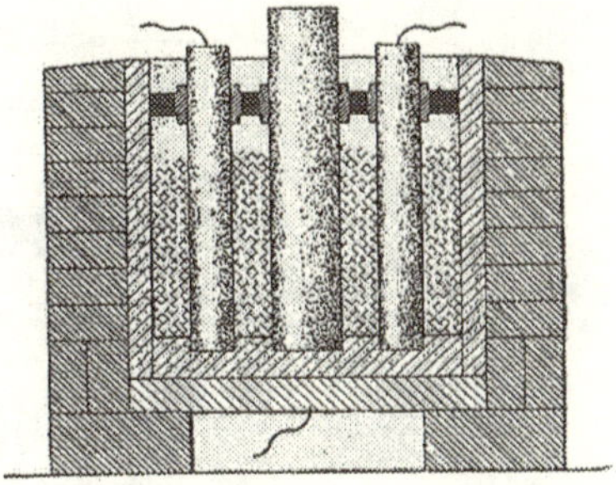

FIG. 101.

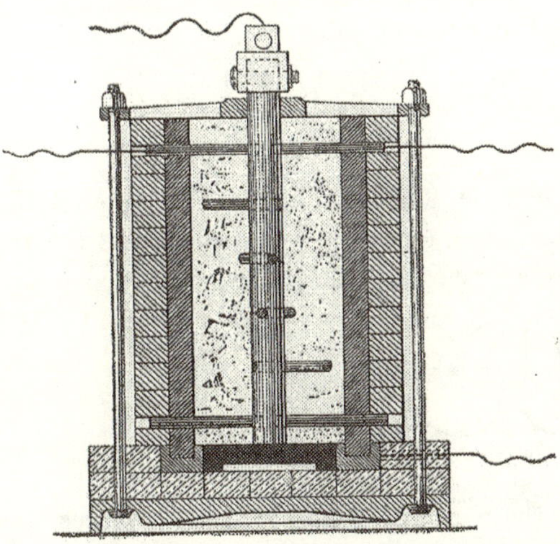

FIG. 102.

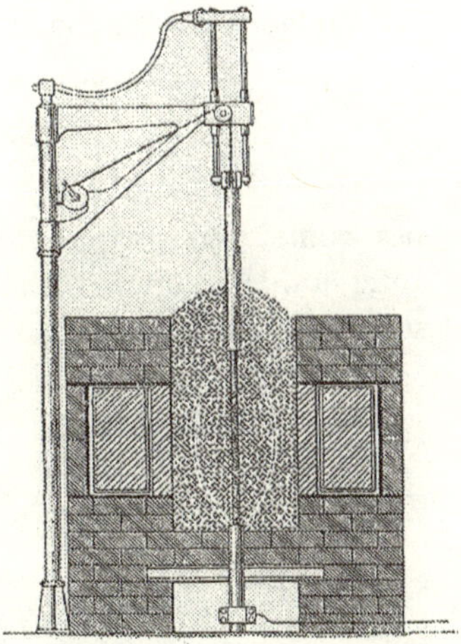

FIG. 103.

are disposed relatively to one another so that they form a skeleton screw-thread as in churning machinery (Figs. 101, 102, 103).

[1] United States Patent, Nos. 577,317, 577,493, and 586,824 of 1897.

Acheson's Core-protected Furnace.—Acheson,[1] in order to protect the glowing carbon core from the action of the materials of the charge, proposes to form a protecting layer of carbide round the carbon resistance during the working. This he

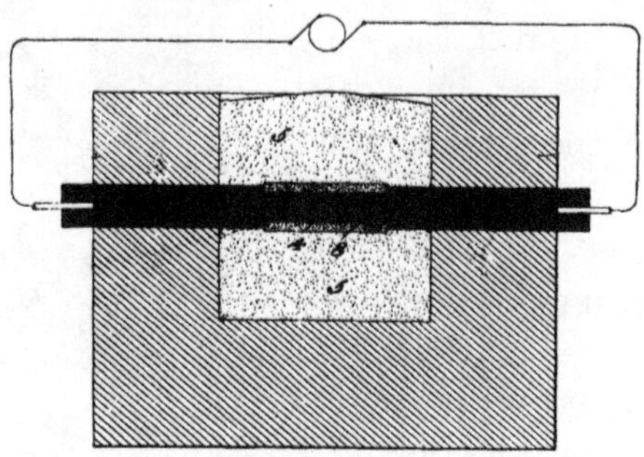

FIG. 104.

accomplishes as follows:—On introducing the charge, he places round the core (4) a loose layer of carbide, *e.g.* calcium carbide, or raw material suitable to its formation, which becomes baked together by the red-hot core (Fig. 104).

Fitzgerald's Furnace.—For heating substances which should

FIG. 105.

not come into immediate contact with the heating resistances, Fitzgerald[2] proposes the arrangement of the latter above the substance to be heated, and shapes the resistances as shown in

FIG. 106.

Figs. 105 to 110 by cutting them out of graphite bars or plates. Resistances of this kind are to be fixed in the manner illustrated in Figs. 109 and 110 beneath covers of charcoal slabs, C. A resistance having the dimensions given in Fig. 107 gives a current

[1] German Patent, No. 159,282, December 10, 1902; [English Patent, No. 27,179, December 9, 1902].

[2] "Electrochemical and Metallurgical Industry," **3**, p. 215 (1905).

of 260 amperes at a pressure of 40 volts applied to its terminals for a length of 500 mm. The charge M is situated beneath this resistance, and is insulated from the carbon contacts T_2 and T_3 by layers of charcoal, B. The charcoal covering C is surmounted

FIG. 107.

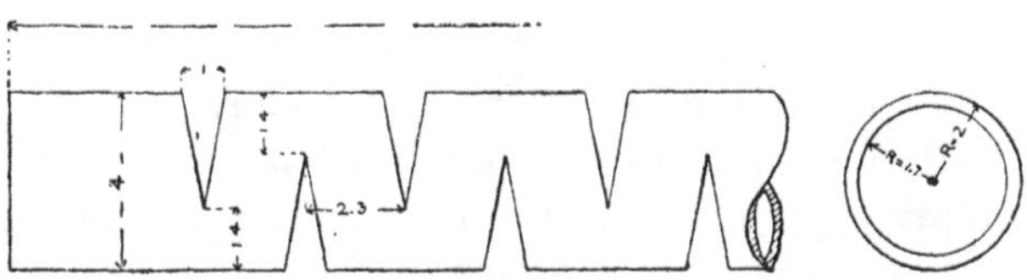

FIG. 108.

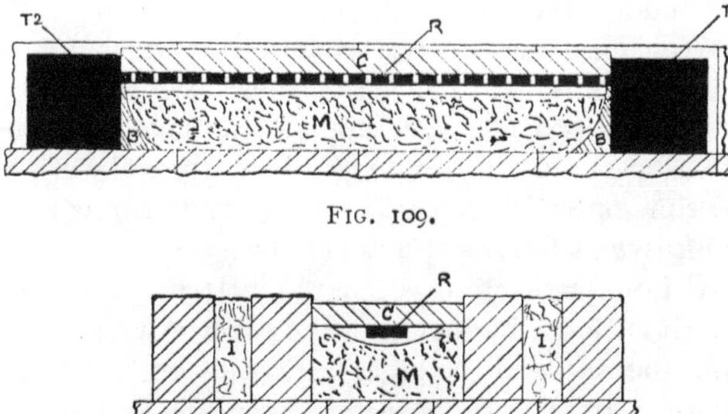

FIG. 109.

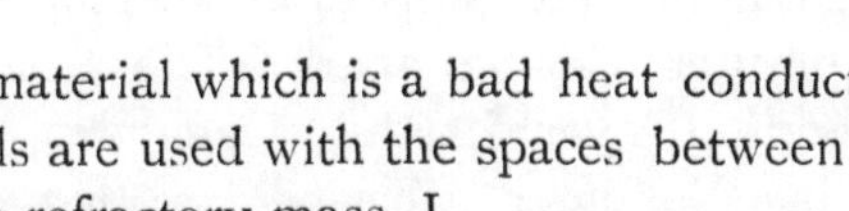

FIG. 110.

by a layer of material which is a bad heat conductor, and double stone side walls are used with the spaces between them likewise filled in with a refractory mass, J.

Closed Electrical Furnaces.—Especially for experimental purposes furnaces of this group have found a very extended application, and as the substance to be heated is no longer in contact with the heating resistance, they may be denoted by the generic name of *closed electrical furnaces*, such as crucible, muffle, tube furnaces with the resistance arranged external to the heating vessel, or heating bodies in which the resistance is enclosed so that they can be dipped into or embedded in the substance to be heated.

Very closely approaching these is the application of metal wires of large heat permanence for winding round the heating vessels named, or for placing them inside the heating bodies.

For experimental purposes in which only moderate temperatures up to about 600° C., at most 700° C., are to be maintained, iron or nickel wire windings, which one can easily make one's self, will suffice.

I, therefore, do not wish to enter more fully into the numerous heating appliances fitted with wire coils which are to be found recorded by hundreds in lists of patents and registered designs, and in the more widely distributed technical literature ; only a few laboratory apparatus, which can be easily made by any metal worker, will be given as examples.

Borchers' Laboratory Furnace. — In the metallurgical laboratory under my direction in the Technical College at Aachen, a calcining furnace was required in which for lengthy periods temperatures had to be kept as nearly constant as possible between 400° and 500° C. We happened to have some porcelain casks such as are used for grinding-mills in dry and wet grinding. In a short time one of these casks was converted into a rotatable calcining furnace as follows (Figs. 111, 112) :—

D is an iron plate attached to the bottom of the porcelain cask *a*, which is closed by means of the cover *b*, while *e* is an iron ring. Both the ring and the plate are rigidly connected together by three iron rods, *e*, *e*', *e*'', in such a manner that these rods do not form any metallic, current-conducting connection between D and *e* ; they are only for the purpose of keeping the plates and the ring at a constant distance apart. The asbestos cylinder *h* forms the supporting layer for the nickel-wire windings, which are protected by the outer asbestos covering *s*. The whole apparatus rests by means of the plate D and the ring *e* on four wheels K, K', $K_{,}$, $K_{,,}$, of which each two are keyed to revolving spindles (*i* and *i*'), which in turn are supported in the bearings *l*, *l*', $l_{,}$, $l_{,,}$. Each of the spindles is made in two halves connected together at the middle by an insulating coupling, so that no current can flow through them. On one of the axles is fixed a grooved wheel, *o*, which acts as the rope-pulley; it is driven by means of a hot-air motor heated by a Bunsen burner.[1]

The electric current, which acts as the source of heat, takes the following path: It enters at one of the bearings, *e.g.* at *n*,

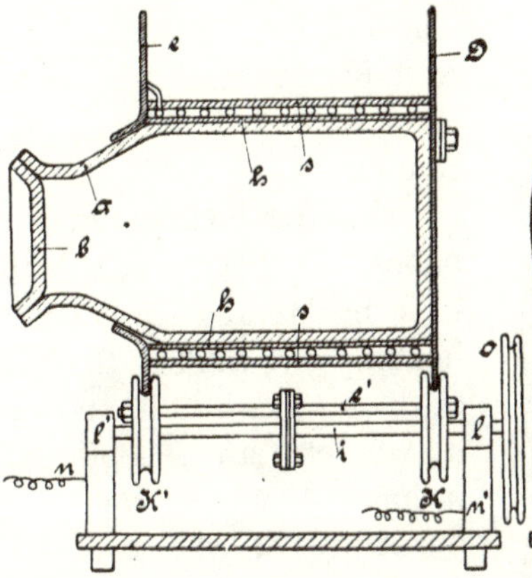

FIG. 111.

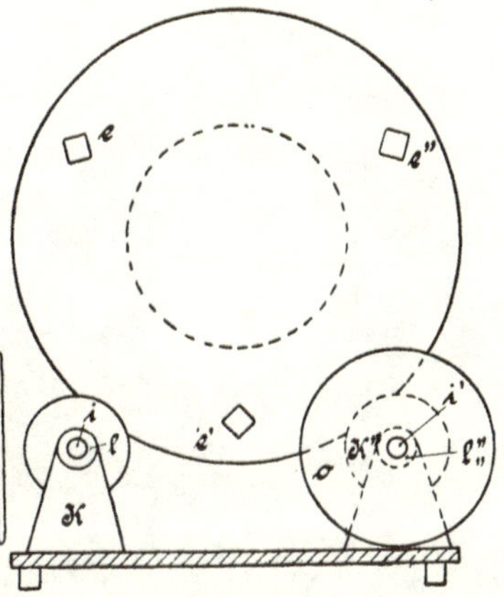

FIG. 112.

goes through the iron wheel *K'* to the ring *e*, traverses the nickel wire, heats the latter, spreads over D, and finally leaves the apparatus at *n'* after flowing through the wheel *K*. The maximum current was 7 amperes at 220 volts. In order to be able to slowly heat up the apparatus and eventually keep the temperature constant at all, a resistance should be included in the circuit. The barrel was rotated at a speed of 8 revolutions per minute, and the temperature could be raised to approximately 500° C. without the porcelain springing.

The Howe Furnace—A specially efficient crucible furnace is the one made by Howe.[2] In Fig. 113 the furnace is shown with the crucible arranged ready for an experiment. The vessel A, in which the crucible F is placed, consists of two half-cylinders made of magnesia with suitable hollow spaces. It can be closed at the top by an annular plate of magnesia B and a plug C provided with a central bore. It has internal spiral grooves in

[1] "Metallurgie," 1904, **1**, p. 292.
[2] H. M. Howe, "Metallurgical Laboratory Notes." New York: 1902.

which a platinum wire or a strip of sheet platinum D is wound, the ends of which, L, K, are led through the side wall and the floor. A tube H is also provided for passing in any gases that may be required. The crucible illustrated here contains metal test-pieces with an enclosed thermo-couple.

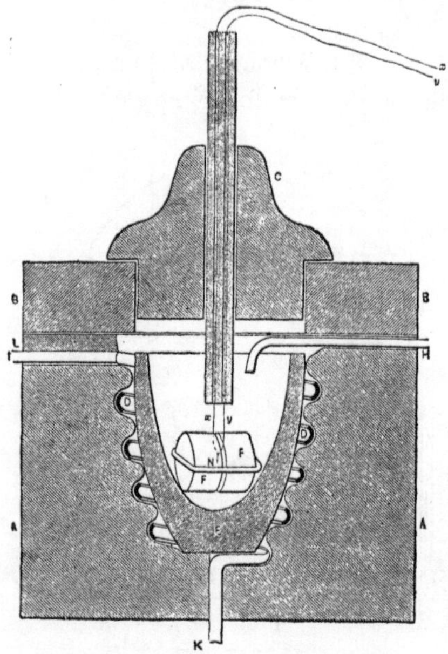

FIG. 113.

The Heräus Crucible Furnaces.—A new crucible furnace by Heräus contains the heating wire built in the body of the furnace and baked with it, whereby an exceptionally intimate contact and high durability of the two, with excellent heat transmission, have been attained. The resistance of such furnaces is subject to very little change; in consequence, the maintenance of uniform temperatures gives rise to very little trouble, an advantage of importance for laboratories, in which a single apparatus cannot be kept under constant supervision.

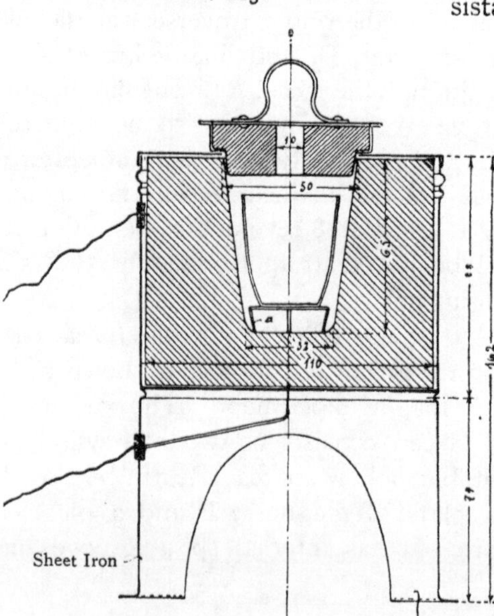

FIG. 114.

Fig. 114 shows a vertical cross-section of this very efficient experimental furnace. The other known "Heräus" furnaces for temperatures up to 1400° C. have porcelain tubes of

20 to 65 mm. clear width, which are spirally wound with platinum foil as the heating resistance, and are enclosed in wider

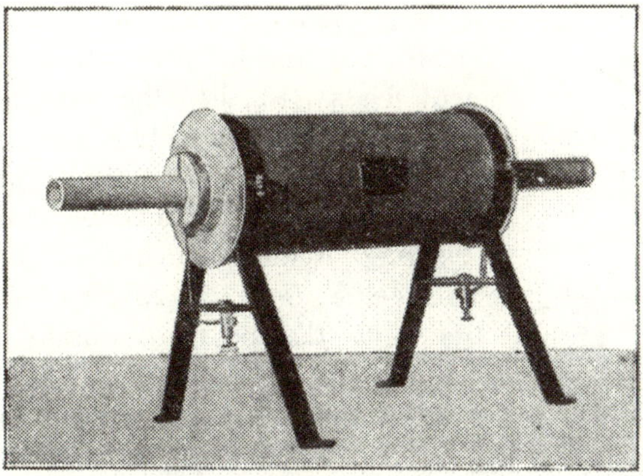

FIG. 115.

FIG. 116.

cylinders filled with refractory material. The furnaces are supplied on stands for horizontal or vertical positions, and can also be had revolvably mounted so that they can be used in any position that may be desired. The use of platinum foil for the heating resistance has very materially reduced the consumption of platinum for resistance heating, in contrast to the furnaces which are worked with wires. Further, the resistances can be easily dimensioned so that the furnaces can be worked with low currents and higher voltages, and can thus be connected to existing lighting circuits of the normal voltages of 65, 110, 220 volts. By altering its width the foil can naturally be wound for any other pressure.

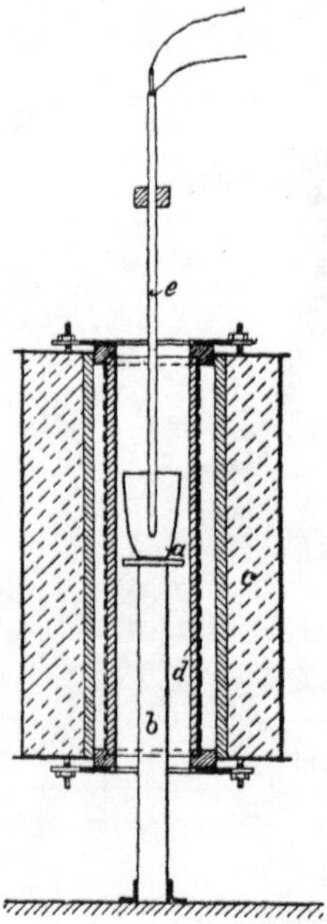

FIG. 117.

With these furnaces the substance under test can be kept under good observation, and the temperature accurately regulated and conveniently measured.

Another considerable advantage of these furnaces is that the tubes enable the air to be excluded either by rarefying it, or by the introduction of other gases.

In Figs. 115 to 117 are given a few types of these furnaces, of which Fig. 117 represents a longitudinal section of one.

The Heräus furnaces give universal satisfaction when they are used for heating to constant temperatures. They are, however, not to be recommended when, for experimental purposes, it is required to quickly produce large temperature fluctuations, as then the porcelain tubes easily spring, and the platinum foil is easily damaged by over-heating, or by the starting of an arc.

Borchers' Granulated Carbon Crucible Furnace.—The usefulness of resistances made of granulated carbon where the preparation of carbonaceous substances connected together presented difficulties, was already pointed out by the Brothers Cowles in their first patents on resistance heating. Acheson, however, was the first to state that resistance substances of definite dimensions could be made in a simple manner by ramming granulated carbon into the charge to be heated. By means of experiments on heating with crucibles which only become conductive at higher temperatures after the manner of the Nernst-glowers, I have used granulated carbon mostly as the preliminary heating resistance, as shown in Figs. 118 and 119.[1] Wedged between

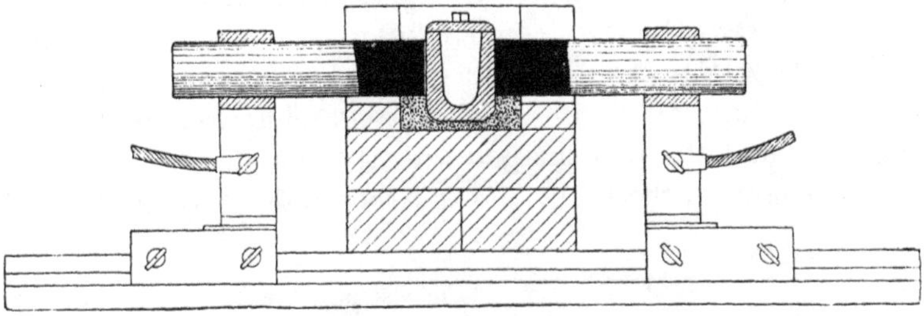

FIG. 118. Scale $\frac{1}{10}$.

two massive carbon rods is a small crucible made of magnesia. It is surrounded with granulated carbon inside a pit made up of a few bricks placed together so that the current—as in the cold state the crucible will not conduct—first traverses the carbon layer, which is thereby made to glow vigorously. As soon as the latter has, however, been raised to a red heat, the crucible itself conducts and becomes connected to the circuit as the resistance material. When it reaches the desired temperature,

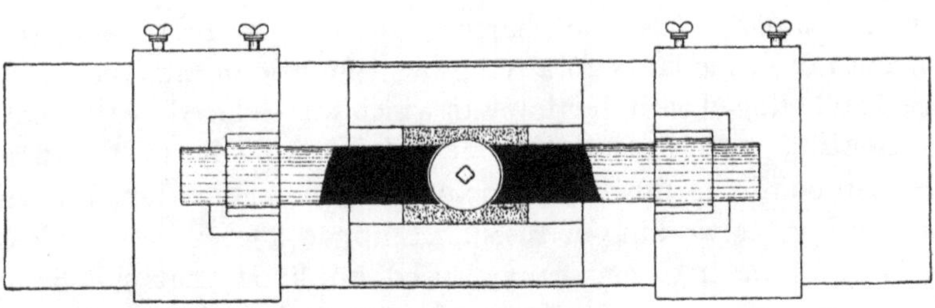

FIG. 119. Scale $\frac{1}{10}$.

the mass to be heated is poured in, the cover is placed on after the first violent reaction ceases, the heating is continued for a little time longer, and then it is allowed to cool. This arrangement should also only be used for experiments on smelting

[1] Borchers, "Das neue Institut für Metallhüttenwesen und Elektrometallurgie an der Königl. techn. Hochschule zu Aachen," p. 38. W. Knapp, Halle a. S., 1903.

which last a short time and with proportionately small quantities of the substance.

The Rossi Furnace.—Already in 1898 Rossi[1] recommended a similar arrangement for melting alloys of titanium. The simple furnace he used (Fig. 120) consisted of a walling C of magnesite with a cover F. The interior of the furnace was filled with powdered charcoal, E, in which the graphite crucible A was placed. The latter was charged with the titaniferous iron mass, and was closed by the lid B. Two carbon rods D passed through

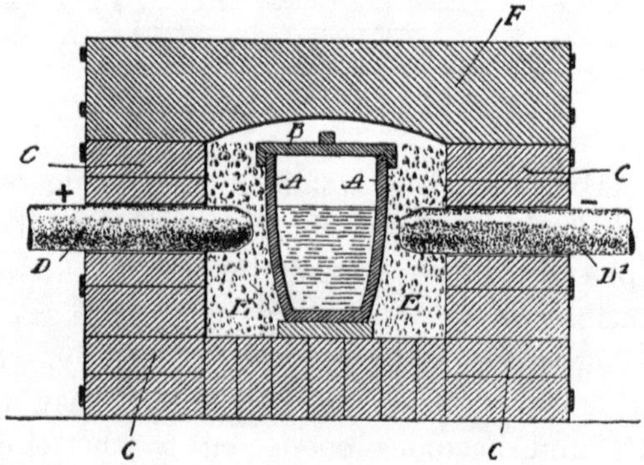

FIG. 120.

the side walls C into the charcoal, and on the passage of the current raised the latter to a red glow ; the iron in the crucible A fused and alloyed with the titanium which was reduced by the heat.

Heating with Granulated Material.—Already in the first German edition of this book I pointed out how important it is, in supervising the working of resistance furnaces when more or less coarsely granulated masses are used, to know that a definite weight of resistance exists in a definite length and for a definite cross-section ; in other words, how important it is in these cases to know the current density in reference to the weight of the conducting material per unit length of the heating resistance.

"Kryptol."—By means of experiments lasting over several years, which were carried out by the Kryptolgesellschaft (Dr. A. Völker), ways and means have been found of not only fulfilling these, but also other, conditions, which have determined a more

[1] United States Patent, Nos. 609,466, 609,467 of August 23, 1898 ; English Patent, No. 18,127 (1898).

extended application of granulated carbon as a resistance material capable of easy handling.

These conditions consist in using the carbon—

1. In forms of perfectly definite conductivity.
2. In mixtures of the one kind with the other in definite proportions.
3. Mixed with non-conducting substances for the preparation of grades of desired electrical conductivity.
4. In uniform grains to be specially chosen for each apparatus and purpose.

The name "kryptol" has been given to this resistance material, which consists principally of carbon admixed with suitably chosen ingredients. It is now supplied in different granulated forms and in different grades of electrical conductivity, and furnaces furnished with it are called "kryptol" furnaces.

"Kryptol" Furnaces.—The furnaces manufactured by the

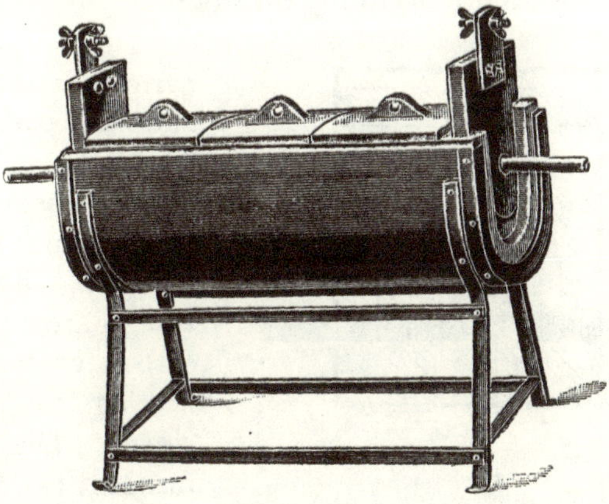

FIG. 121.

Kryptolgesellschaft[1] are so arranged that by loosely filling in the furnace space provided with a definite kind of kryptol (avoiding the formation of hollow spaces), a fairly exact resistance will be attained. Any one can, therefore, in a short time himself apply and change the heating resistance for the heating vessel (crucible, tube, muffle, etc.).

In order to demonstrate the many-sided application of heating

[1] Supplied in this country by W. F. Dennis & Co., London.

with kryptol, I give the following illustrations of some furnaces supplied to me by the Kryptolgesellschaft for the Aachen

FIG. 122.

Metallurgical Institute, as well as some furnaces constructed for special purposes according to my directions.

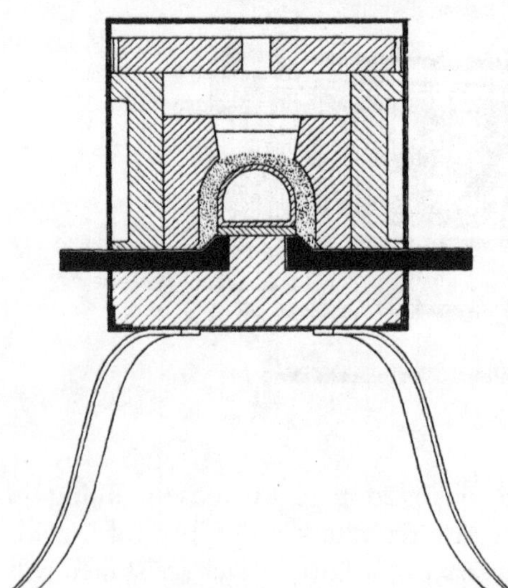

FIG. 123. Scale $\frac{1}{8}$.

Tube furnaces for narrower and wider glass and porcelain tubes:—The tube to be heated rests in a semi-cylindrical trough which is filled with kryptol. The ends of the trough are provided with carbon contacts for connecting to the supply circuit. The longer tube furnaces are further provided with adjustable contacts by means of which any lengths of the furnace can by short-circuiting, or the inclusion of resistances, be wholly, or partially, cut out of circuit, and so less strongly heated than the rest of the furnace-trough together with its contents (Figs. 121 and 122).

A muffle furnace has the arrangement shown in Fig. 123.

A crucible furnace is illustrated in Fig. 124. Both sketches do not require any further explanation.

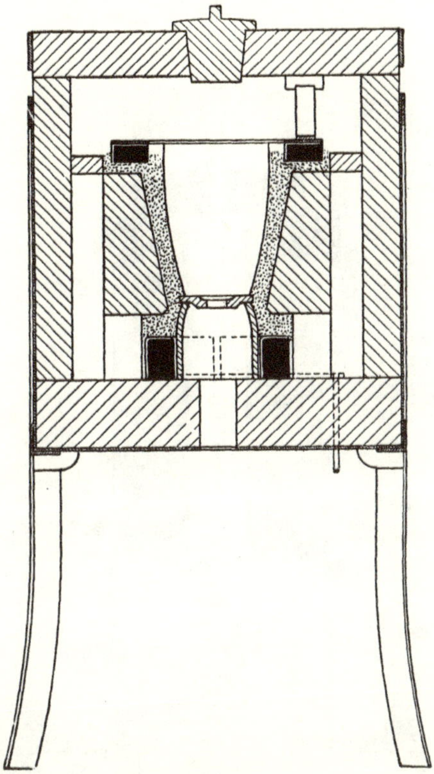

FIG. 124. Scale 1/10.

"Kryptol" Furnace of Borchers and Glaser.—For calorimetric investigations F. Glaser[1] constructed, in accordance with my directions, the kryptol furnace depicted in Fig. 125, together with the very simply made calorimeter. As will be seen, it is constructed in an iron box standing on feet. On an annular base of fire-brick, having an inside diameter of 70 to 65 mm., rests a carbon ring which forms one of the current conductors and encloses the lower extremity of a highly refractory tube. The latter has a clear internal width of 80 mm., a wall thickness of 5 mm., and a length of 400 mm. On the carbon ring stands a hollow cylindrical shaft made of fire-brick, leaving a free space of 20 mm. between it and the tube. The continuation of the shaft is formed by a rectangular carbon rim which serves for the other conductor and slightly extends above the inner tube. The space between the rim and the shaft on the one hand and the tube on the other was carefully filled in with small fragments of retort-carbon about 2 mm. thick. The whole apparatus is closed at the top by two carbon blocks, which are provided with a boring through which the thermo-couple enclosed in its cover can be inserted. The heating of the tube is effected by sending a current of the requisite intensity through the carbon fragments.

[1] "Metallurgie," 1, pp. 105–107 (1904).

The inner cylinder is closed at the bottom by a carbon block which just fits the opening of the lower fire-brick ring, and is screwed to an iron flap. The flap is kept shut during the experiment by means of a piece of flat iron bar which is in turn held by two lugs, of which the one is visible in the drawing beneath the iron case. By drawing back the flat iron bar, the flap drops, and the heated substance falls into the water in the calorimeter. When experiments are to be made with fused masses, it is further necessary to stop up the joints between the carbon and chamotte ring with asbestos fibre. The metal can then be at once melted without any risk of its running out. The calorimeter consists of a sheet-iron box having the dimensions 350 × 350 × 400 mm., and stands on rubber feet in a wooden case. The space between the iron box and the case is filled in with wood-wool. Two stirrers are arranged in opposite corners, and are vigorously rotated in opposite directions by a small hot-air motor. For

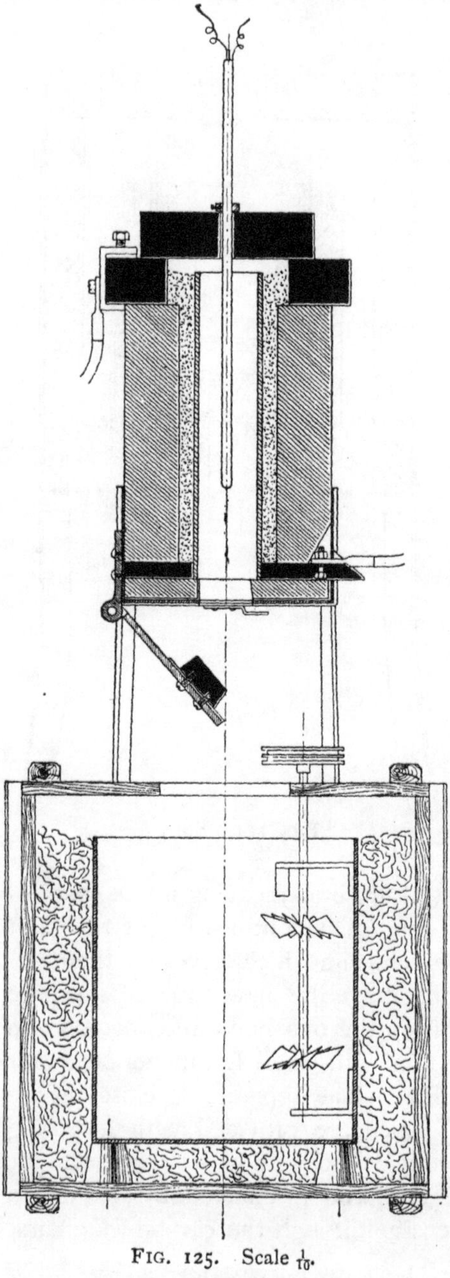

FIG. 125. Scale $\frac{1}{10}$.

the temperature determinations, a Beckmann thermometer and a standard thermometer with the degrees divided to tenths are used and are read with a magnifying glass.

Hesse's Furnace.—With a different arrangement, R. Hesse utilizes the apparatus for heating samples of material in a current of nitrogen. In its essential details the arrangement of the apparatus remains the same as with the Glaser experiments (Figs. 126 and 127).

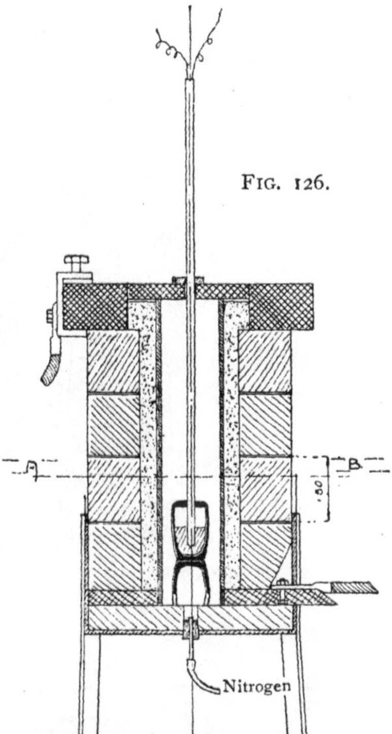

FIG. 126.

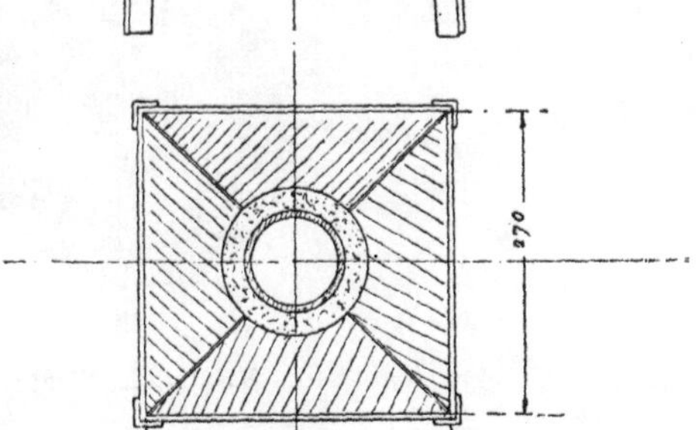

Section at A B

FIG. 127. Scale 1/10.

Borchers' Carbon Crucible "Kryptol" Furnace.—As already pointed out by the Kryptolgesellschaft in their publications, the attainable temperature is only limited by the resistance capacity of the heating vessel used. When vessels of carbon can be selected, I should like to add that the upper temperature limit is at the temperature of volatilization of the carbon. Where high temperatures inadmissible with porcelain and clay vessels were required, I adopted the following arrangement. I formed the heating space inside one of the carbon poles, which consisted of a carbon rod hollowed out at the top. This carbon pole stands in the centre of a very much larger pole, also of carbon (cp. the calcium apparatus, p. 45, and Figs. 49 and 50). In this way very heavy currents can be sent through the kryptol mass from the large outer carbon pole to the small inner one. The current stream-lines, and therefore the heat effect, are thus con-

FIG. 128.

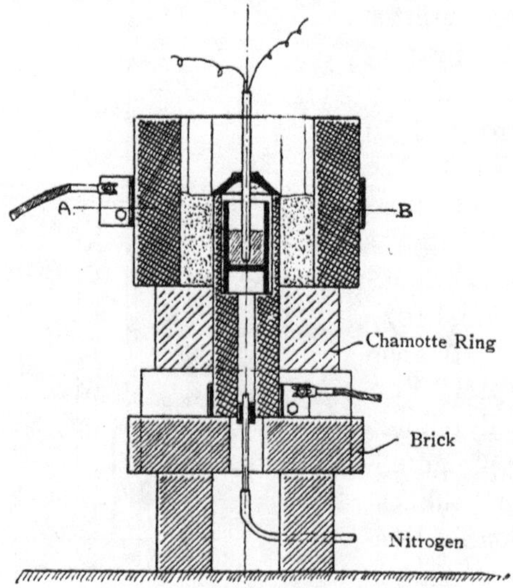

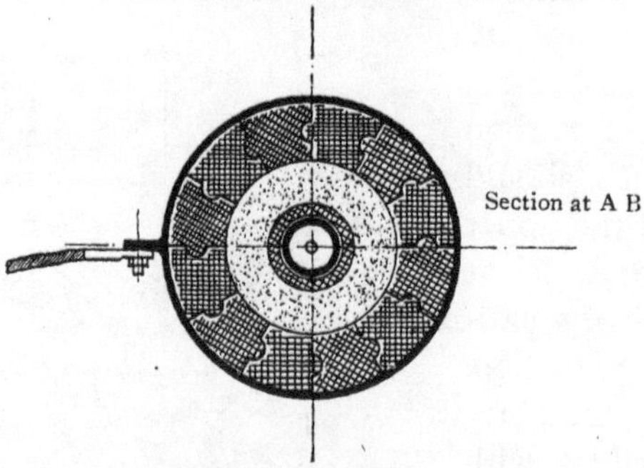

FIG. 129. Scale $\frac{1}{10}$.

centrated from the cylindrical envelope towards the axially-placed crucible in proportion to the concentric circles of the inner shell of the furnace and the outer periphery of the crucible. By making the thickness of the wall of the crucible, formed by boring out the middle carbon rod, as small as possible, the resistance heating can be still further increased at this point (Figs. 128 and 129).

CHAPTER IV

DIRECT ARC HEATING

FURNACES IN WHICH THE SUBSTANCE TO BE HEATED FORMS ONE OR BOTH POLES OF AN ARC

The Davy Arc Furnace.—It is not improbable that Davy[1] already, in 1810, in his experiments on the electrolysis of alumina, mentioned on p. 5, worked for a short time with the electric arc, and therefore used the first direct arc furnace of the smallest size for the extraction of metals.

The Despretz Arc Furnace.—The next experimental furnace was made by Despretz. No doubt as a preliminary experiment for his investigations on the action of high degrees of heat on diamonds.

Despretz, in the year 1849, tested the behaviour of a small retort, made of sugar-carbon (15 mm. diameter), at the temperature of the arc which he allowed to form inside the retort between the retort itself, as the positive pole, and a pointed carbon rod. From the sugar-carbon graphite was formed in places mostly exposed to the action of the arc.

The Siemens Arc Furnace.—With the sources of current at that time available, it is easily understood that this principle of heating could not be developed in a manner beneficial to the technical industry. It is, therefore, not to be wondered at if new small arc furnaces were not introduced until the years 1878 and 1879. Even then they were extremely perfect; they are the furnace constructions of Charles W. Siemens.

One of these was arranged as follows:—Through the bottom of a crucible, enclosed in a metal shell, a conducting pole was

[1] *Comptes Rendus*, **29** (1849).

taken into the interior of the crucible, so that its end comes into direct contact with the molten substance. The end of the pole was, however, protected by a platinum tip, or other resisting material, so that the substances to be fused were not contaminated by it. The other pole was passed through the lid into the interior of the crucible, and consisted of a hollow metal body, which, as shown in Fig. 130, could be cooled with water or other liquids. Later on the poles were interchanged, as will be seen from Fig. 131; the cooled pole was transferred to the floor of the crucible, while the pole hanging through the lid consisted of carbon.[1]

According to a communication by Siemens in the year 1880, the arrangement and method of working with the altered furnace were as follows:—

"A melting-pot of graphite or other difficultly fusible material

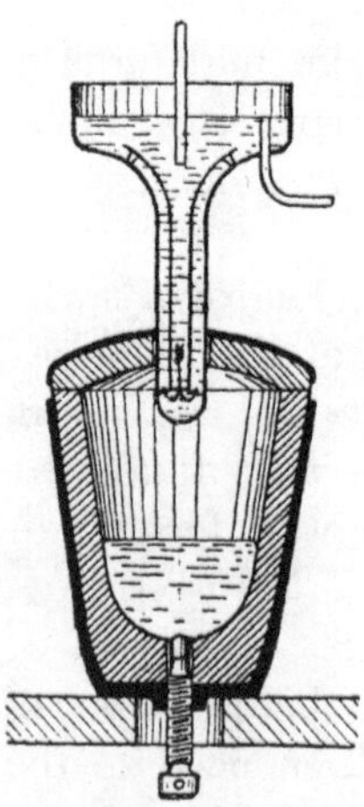

FIG. 130.

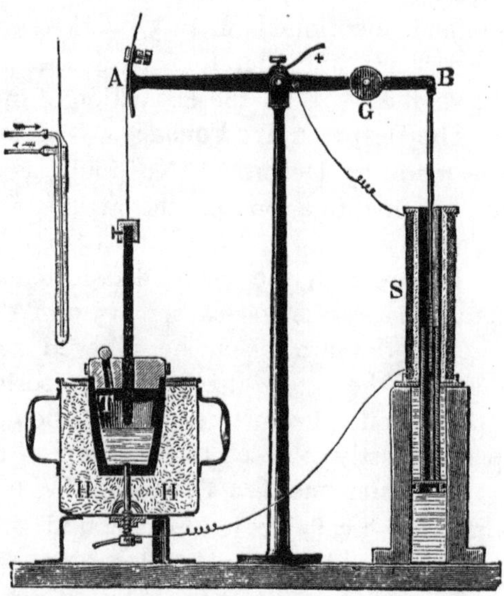

FIG. 131.

is placed in a metal vessel, the space between them being filled in with pounded charcoal or other bad heat conductor. Through the perforated floor of the crucible a rod of iron, platinum, or gas

[1] English Patents, No. 4208 of 1878 and No. 2110 of 1879.

carbon, such as is used in electric lighting, is introduced. The likewise perforated crucible cover receives the negative electrode. If possible, the latter is made in the form of a cylinder of compressed carbon of comparatively large dimensions. The negative electrode is suspended from the end A of a centrally supported beam AB by means of a band made of copper or other good conductor, whereas to the other end of the beam is attached a hollow cylinder of soft iron, which can move freely in the vertical direction inside a wire coil S, having a total resistance of about 50 ohmic elements. The magnetic force with which the hollow iron cylinder is sucked into the solenoid is balanced by means of a running or sliding weight G on the arm of the beam nearest the wire coil. By altering the position of this weight the resistance of the arc can be determined, and fixed within the limits of the source of power. The one end of the wire coil is connected to the positive, and the other end to the negative pole of the electric arc. With increased resistance of the arc the negative electrode dips deeper into the smelting crucible, while with decreased resistance the weight forces the iron cylinder down into the coil, whereby the length of the arc is increased until the balance between the acting forces is again restored. This self-regulating action is of great importance, as, on the one hand, without it the temperature in the melting crucible would decrease, and, on the other hand, the sudden decrease of the electrical resistance of the material which is being fused would cause a sudden increase of the resistance of the arc and its probable extinction. For the success of electric smelting it is, further, of importance that the material to be fused should form the positive pole of the electric arc, as it is well known that it is at the positive pole that heat is produced. This arrangement is obviously only applicable with metals; with non-conducting earths, or in the case of gases, it is necessary to provide an indestructible positive pole, which, however, can be subjected to the smelting process, and can form a small pool on the floor of the smelting crucible. The accumulation of heat takes place very rapidly. Using a moderate-sized dynamo-machine of 36 Webers, a smelting crucible of about 20 cm. in depth placed in non-conducting material is brought to a white heat in less than a quarter of an hour, so that 1 kg. (2·2 lb.) of steel is fused in a further quarter of an hour. To prevent the consumption of the negative pole, the author uses a water-cooled pole, or copper pipe, through which a cooling stream of water is allowed to flow. It is a simple copper cylinder closed at the lower end, and contains an inner rubber tube, which very nearly reaches to the bottom, and serves to introduce the jet of water."

According to the calculations of Siemens, with a steam-driven dynamo one pound of coal will, theoretically, nearly fuse one pound of cast steel.

"An ordinary blast furnace consumes from 2½ to 3 tons of best coke in fusing a ton of steel in smelting crucibles, a regenerative furnace only 1 ton, while in smelting on the open hearth of the same furnace 12 cwt. of coal are consumed. The electric smelting furnace, therefore, comes equal to the regenerative gas furnace as regards fuel economy. In addition, the following advantages are in favour of the process:—1. That the attainable degree of heat is, theoretically, unlimited. 2. That the melting takes place in a perfectly neutral atmosphere. 3. That the process can be undertaken without much preliminary preparation, and under the eye of the observer. 4. That by using the ordinary difficultly fusible materials the limit of heat practically attainable is very high, as in electric furnaces the material to be molten has a higher temperature than the melting crucible itself, whereas in the ordinary process the temperature of the melting crucible exceeds that of the material fused in it. Even if the electric furnace will not supplant the ordinary smelting furnaces, still by its aid chemical reactions of the most varied kind will in future be capable of being carried out, and at temperatures which, up to the present, were impossible.

"From work carried out with this apparatus the following examples may be cited:—

"One crucible charge of 10 kg. (22 lb.) of steel was completely fused in one hour.

"For liquefying 4 kg. (8·8 lb.) of platinum one quarter of an hour sufficed.

"A smelting experiment with copper, which was packed in coal-dust, yielded an evaporation loss of more than 90 per cent."

I must specially draw attention here to an important point in the Siemen's patent specification which has been overlooked in most of the reports on these furnaces, namely, to the attempt to confine the arc electro-magnetically to a definite position. In the English patent specification, No. 2110 of 1879, it is recommended *to surround the crucible with a wire solenoid traversed by a current in order to counteract the tendency of the arc to spring across to the crucible walls.*

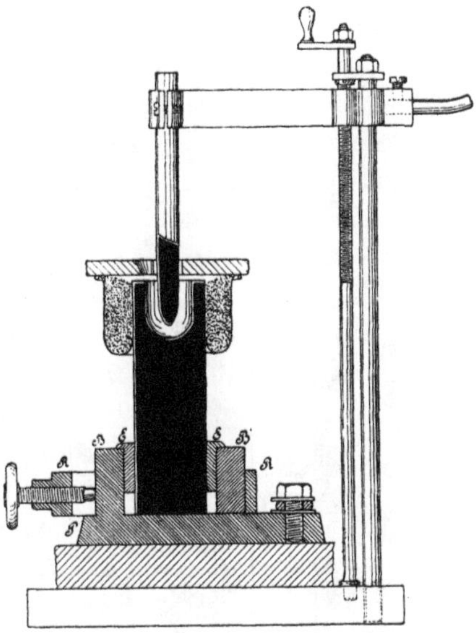

FIG. 132.

Borchers' Direct Arc Furnace. — One of the smallest experimental furnaces which can be worked with the small currents of house and laboratory lighting installations I had constructed in 1892 for lecture and laboratory experiments.

The upper end of a long, thick carbon rod is bored out to form the interior of the crucible (Figs. 132 and 133). The lower extremity of this carbon rod is placed in a holder which at the same time enables the connection with the current circuit to be made. The holder consists of a base-plate P, a fixed cheek B cast with it, and a cheek B′, which is loosely placed on P. By means of the ring R, provided with a screw, the loose cheek can be drawn up tight against the first one. When it becomes necessary to use different sizes of crucibles, inset blocks E are placed between the cheeks which clamp the crucible ; and for each size of crucible one pair of blocks E must be kept in readiness. In order to prevent the carbons from burning away in the region of the melting zone during experiments on smelting, a sheet-iron box, which should be filled with charcoal, is pushed over the upper edge of the crucible.

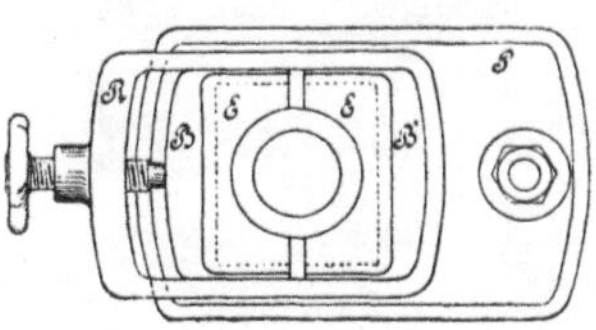

FIG. 133.

The 300-Ampere Arc Furnace of the Deutsche Gold- und Silberscheideanstalt.—Since 1895 a furnace of the following construction (Fig. 134) has been supplied by the German Gold and Silver Refining Institution for currents from 100 to 300 amperes:—According to the purpose for which it is intended, either a carbon melting crucible or one composed of sintered magnesite is made use of. In the first case the crucible also acts as the negative electrode. When it consists of non-conducting material, the current is conducted through a hole in the crucible floor by means of a carbon pin. The positive carbon, which is introduced from above, can be raised or lowered by means of an adjusting screw according to the requirements and to the source of current at hand. A screw terminal provided on the holder serves for clamping the positive cable.

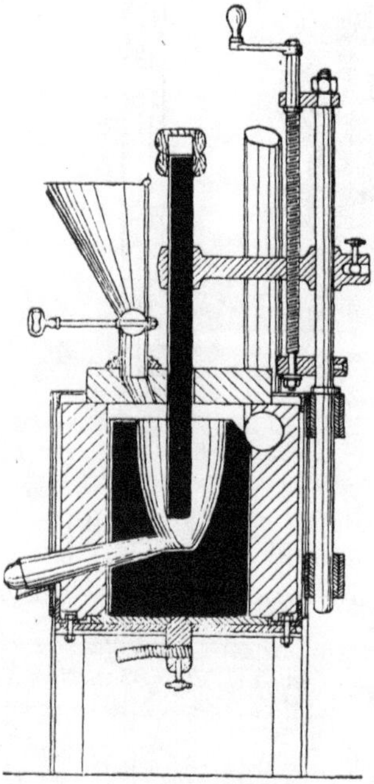

FIG. 134.

When material in the form of powder is to be worked in the furnace, the raw material can be constantly added by means of a feeding hopper—for the manufacture of calcium carbide, *e.g.*, the mixture of chalk and coal—and the molten mass can be run off by means of the tapping hole in the crucible floor. Any gases evolved during the process escape through the outlet pipe inserted in the wall.

The Arc Furnaces of Thwaite and Allen.—Especially for smelting metals, Thwaite and Allen furnish their graphite crucibles with the following electrical smelting arrangements (Figs. 135 to 137):—The metal B, Fig. 135, contained in the clay or graphite crucible A, is kept away from the carbon pole E by the tube D consisting of fire-proof material, so that the arc is formed deep down in the interior of the mass to be fused. The molten metal C accumulates on the floor of the crucible. By aid of the lid G, in which the tube D is clamped, and the guide bars

F, which are in connection with wire or cable terminals, the carbon E is placed in circuit, while on the other side the connection with the current source of the material to be melted is established by means of the metal strips H. These are suspended in a sheet metal ring, placed in the upper part of the crucible, with its lower edge bent up inwards. Finally, the sheet ring can be connected to the circuit through a screw terminal. The actual holder for the carbon E is formed by the split tube J provided with the handle K.

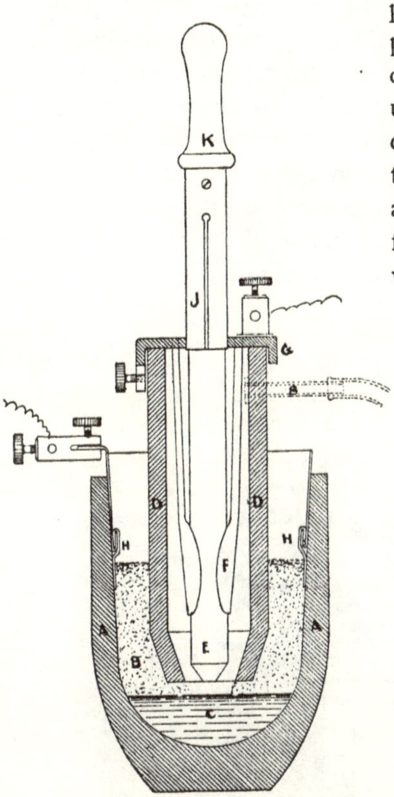

FIG. 135.

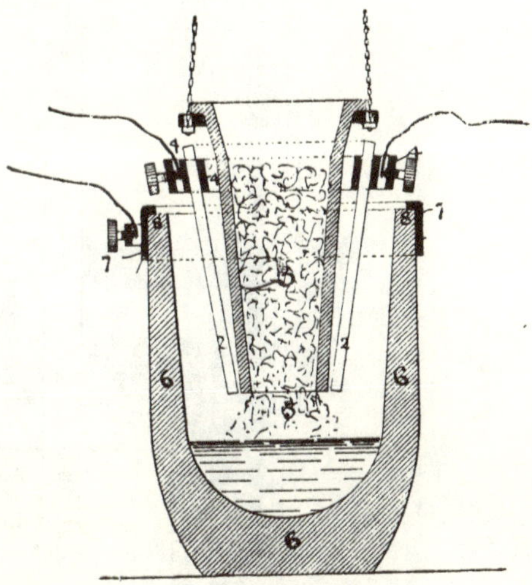

FIG. 136.

Another form of this smelting crucible is represented in Figs. 136 and 137. Round a funnel 3, in which the metal to be fused 5 is introduced, are arranged the carbon rods 2 clamped in a ring 4, which is connected to the circuit. By means of the crucible, which is composed of conducting material and is placed in circuit through a metal ring 7 on its rim 8, the metal to be fused forms the second pole of the arc, which jumps across from the carbon rods 2.

The 1000-Ampere Arc Furnace of the Deutsche Gold- und

Silberscheideanstalt.—In a larger experimental furnace of the German Gold and Silver Refining Institution for currents of about 1000 amperes, the heating chamber is lined with magnesite masonry, or only with magnesite at the top and graphite at the bottom, G and C (Fig. 138). The furnace chamber is heated by the electric arc. The current is conducted by means of the thick carbon rods, of which the negative carbon D enters from below and is fixed, whereas the positive carbon D_1 enters from above through the lid H, and can be moved up and down by the

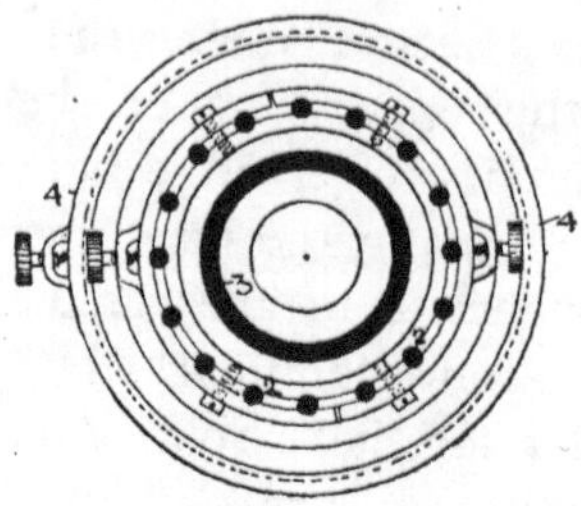

FIG. 137.

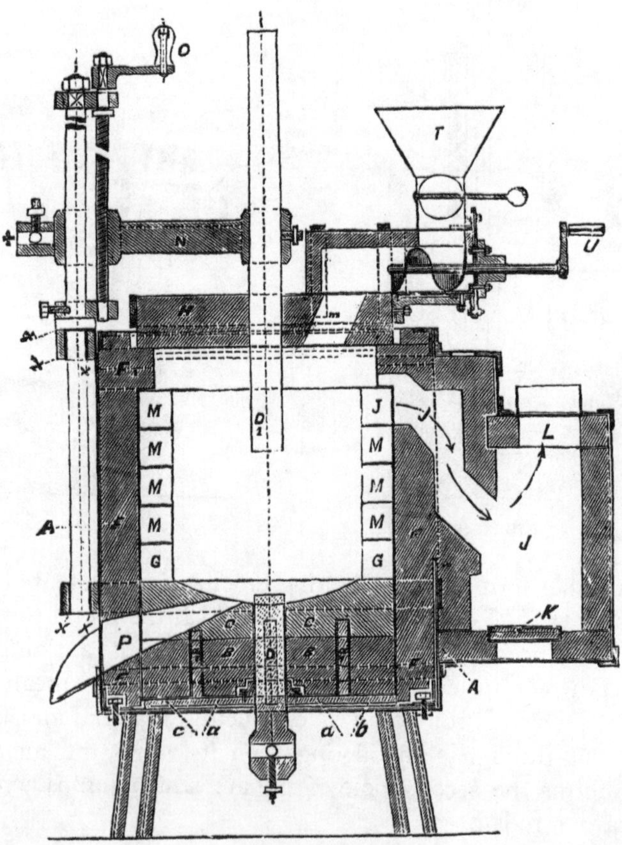

FIG. 138. Scale $\frac{1}{20}$.

mechanism NO provided with a handle. In this manner the length of the arc is varied according to the pressure at one's disposal. At the commencement of the intended process it is advisable to short-circuit the carbons D and D_1 by means of thin carbon pencils provided for this purpose, so as not to get too high a temperature momentarily, as this would be prejudicial to the life of the carbons. The furnace is charged through the hopper T. The gases evolved during the process escape in the direction of the arrows through the channel J, and pass through the opening L into the atmosphere, or into the flue. Any solid particles carried away with the gases settle in the channel J, and can be removed by drawing out the slide K. The materials melted in the crucible are run off at the tap-hole P. During the operation of smelting the tap-hole is kept closed by a carbon plug. The furnace stands on four feet. To be better able to renew the negative carbons, it is advisable to prolong the feet through brick foundations. The furnace floor is supported by the plate *b*. On this plate *b* is an asbestos plate *c*, about 9 mm. thick, and on this follows the thick iron plate *a*. In this iron plate *a* the four iron bolts a_1 are let in, and they reach up into the graphite floor C. The object of this arrangement is to conduct the current from the negative electrode without large voltage losses. The controlling mechanism is insulated at the supports x, x, x, etc.

Slavianoff's Arc Furnace.—For fusing metals specially for the purpose of alloys Slavianoff's crucible should be mentioned. It differs from all smelting devices hitherto made known in so far as both the poles of the arc are formed by the material to be melted. The metal to be liquefied is arranged as the upper pole in the form of a rod, while the fused metal itself forms the negative pole (Fig. 139). For foundry purposes, or for metal refining, the absorption of carbon by the metal under fusion, unavoidable with the other

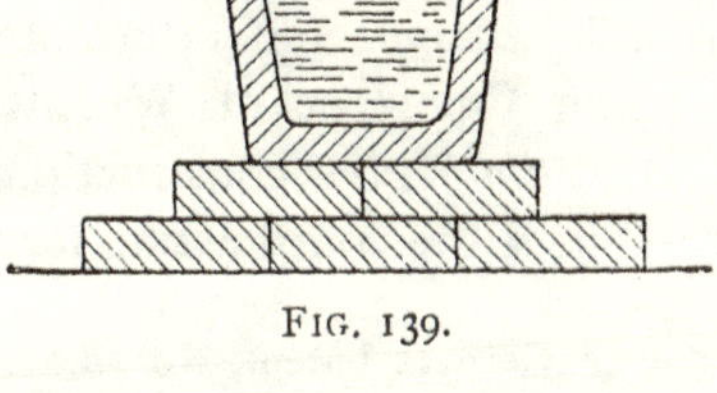

FIG. 139.

apparatus, is completely frustrated. It is, of course, clear that, in consequence of the metal dropping off the top electrode, some difficulties occur in the maintenance of the arc. Slavianoff has, however, removed these by using a sensitive regulating device.

The Arc Furnace of Gérard-Lescuyer.—The idea to form the arc from the materials of the furnace charge has, moreover, already been given expression to in older publications. Gérard-

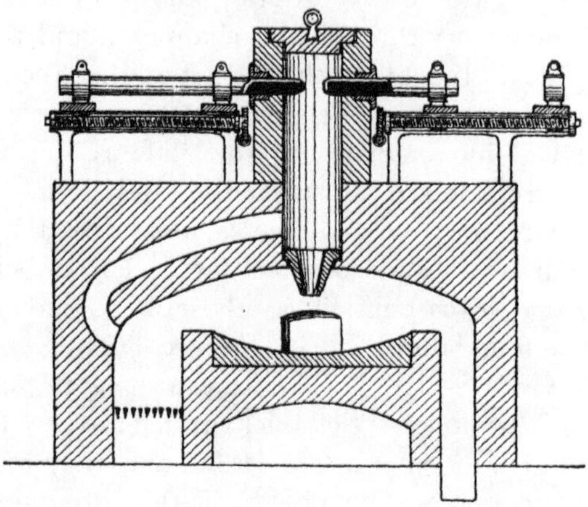

FIG. 140

Lescuyer (Fig. 140)[1] proposed to manufacture aluminium-bronze, for instance, by preparing rods of aluminium oxide, carbon, and copper, and by letting the arc form between them. The metal as it drops down was to collect on the hearth of a reverberatory furnace, which was kept hot by means of a small fire, in which also the carbonic oxide formed during the process of reduction was burned. Even if this apparatus should prove of use at all, it certainly was not adapted for the production of an alloy both the components of which are extremely easily volatile at the temperature of the arc.

Arc Furnaces of Whitney and Strong.—Later Whitney[2] and Strong[3] give constructions of furnaces with which are also combined the compressors for the electrodes, the latter composed

[1] German Patent, No. 48,040; [English Patent, No. 18,994 (1888)].

[2] United States Patent, No. 578,685.

[3] United States Patent, No. 587,343.

of the materials of the charge. Whitney uses conical tubes with a spiral compressor, Strong brick-mills, with rollers and pistons. The compressed bodies are to be baked by the heat which escapes outwards from the furnaces.

The Willson Furnace.—It is almost unnecessary for me to add that the furnaces which work with carbon crucibles, or carbon linings, and with carbon anodes suspended from above, even when in these instances they utilize the electric arc for heating purposes, are more or less exact copies of the first Héroult furnace (p. 11). Even the so-called Willson furnace, with which Willson[1] attempted during the years 1890-94 to manufacture the earth and earth alkali metals by direct reduction of the oxides with excess of carbon, but finally obtained their carbides, is in all its essential parts an unchanged Héroult furnace (Fig. 141).

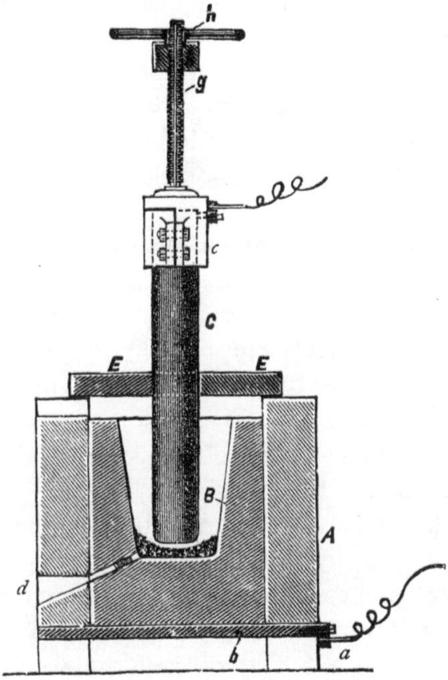

FIG. 141.

A carbon crucible B is fixed in the wall of masonry A, and is connected to the circuit at *a* through the metal plate *b*. A carbon rod C, which is suspended by means of the clamping-piece *c* to a screw spindle *g*, which can be moved up and down by the handwheel *h*, forms the other pole. As much coal is admixed with the material it is desired to reduce so that it does not get fused. The inventor, however, expects that the reduced metal would separate out from this carbon mass, so that it could from time to time be run off at the tapping hole *d*.

[1] United States Patent, No. 430,453 of June 17, 1890; No. 492,377 of February 21, 1893. English Patent, Nos. 4757 (1891), 21,701 (1892).

Energy Consumption in Calcium Carbide Manufacture.—After the work done by Moissan on calcium carbide became known in the years 1894 and 1895, reports on the Willson experiments,[1] containing some strongly optimistic exaggerations, were published by the adherents of Willson in all the technical journals and even daily papers. The assertion was then made that with less than 2 k.w.-hours 1 kg. of calcium carbide could be produced, whereas from my experiments I had determined the energy consumption to be at least 4 to 4·3 k.w.-hours, a result which was later completely confirmed in practice on the large scale; in fact, at first by de Korda, who, in 1896, before the Second International Congress for Applied Chemistry in Brussels, based his estimate of the cost on a consumption of 4·4 k.w.-hours per kg. of carbide taken from actual working results; later by Keller, who placed a very interesting summary of the power consumption under different working conditions with currents of 7500 amperes at a mean pressure of 30 volts before the Third International Acetylene Congress in Paris in 1900. These observations are graphically represented in Fig. 142. In the diagram the vertical ordinates on the left give the weights of carbide in kg. per 24 k.w.-hours, and on the right the yield in litres of acetylene of 1 kg. of the corresponding product of the smelting at a pressure of 760 mm. and a temperature of 15° C. (Fig. 142).

The charge itself consisted of lime, of which the amounts of carbon added per 100 kg. are given by the abscissæ in the diagram. An examination of the curves shows that a mixture of 100 kg. of lime with 65 kg. of coal yielded 6·2 kg. of calcium carbide per 24 k.w.-hours with an evolution of acetylene amounting to 300 litres, whereas with a mixture of 100 kg. of lime with 60 kg. of coal the yield was 7 kg. of carbide for the same consumption of energy with an amount of acetylene of 290 litres per kg. of carbide. With both of these very favourable results it would naturally be most advantageous for carbide factories, if commercially one would be satisfied with a yield of 290 litres of acetylene per kg. of carbide. The electrical energy can then be most efficiently utilized.

[1] "Jahrbuch der Elektrochemie," 1904, **1**, 237. W. Knapp, Halle a. S.

Even less favourable costs obtained for the manufacture of carbide were on the whole sufficiently satisfactory to specially encourage those in possession of water-power to very speedily complete their power plants for carbide manufacture. This naturally resulted in very active rivalry in the construction of new electrical furnaces. Most of these designs, on account of the conductivity of calcium carbide, belonged to this category of

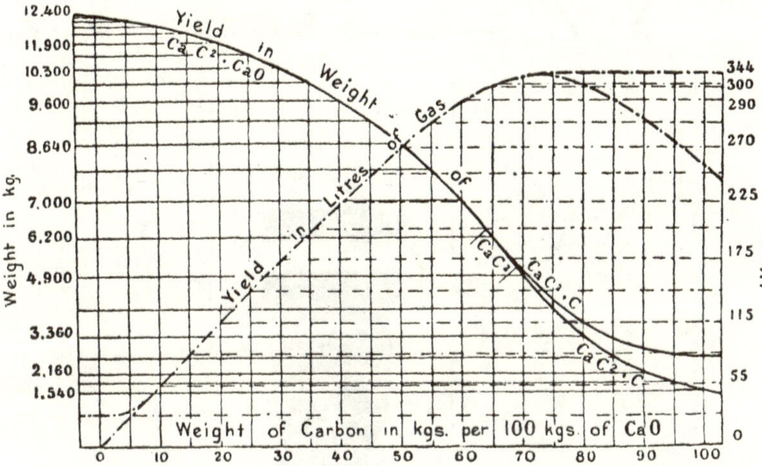

FIG. 142.

furnaces and that of direct resistance heating, or a combination of both. The contest led to very ingenious but also to very complicated constructions. To-day one has again returned to the simplest as the best.

Arc Furnace of the Willson Aluminium Company.—Instead of the furnace represented in the Willson patent specifications, according to the reports in the *Electrical World* (New York, 1895), portable Héroult crucibles were used in the first large carbide installation of the Willson Aluminium Company at Spray, N.C., from which, however, the carbide was not tapped, but tipped out after solidification. At the start of the operation the crucible floors were lined with layers of carbon, while the walls were protected during the working by the portion of the charge which remains undecomposed (Fig. 143).

Arc Furnace of the Société des Carbures Métalliques.—

The so-called block-smelting process, as it has been carried out at the works of the Willson Aluminium Company at Spray,

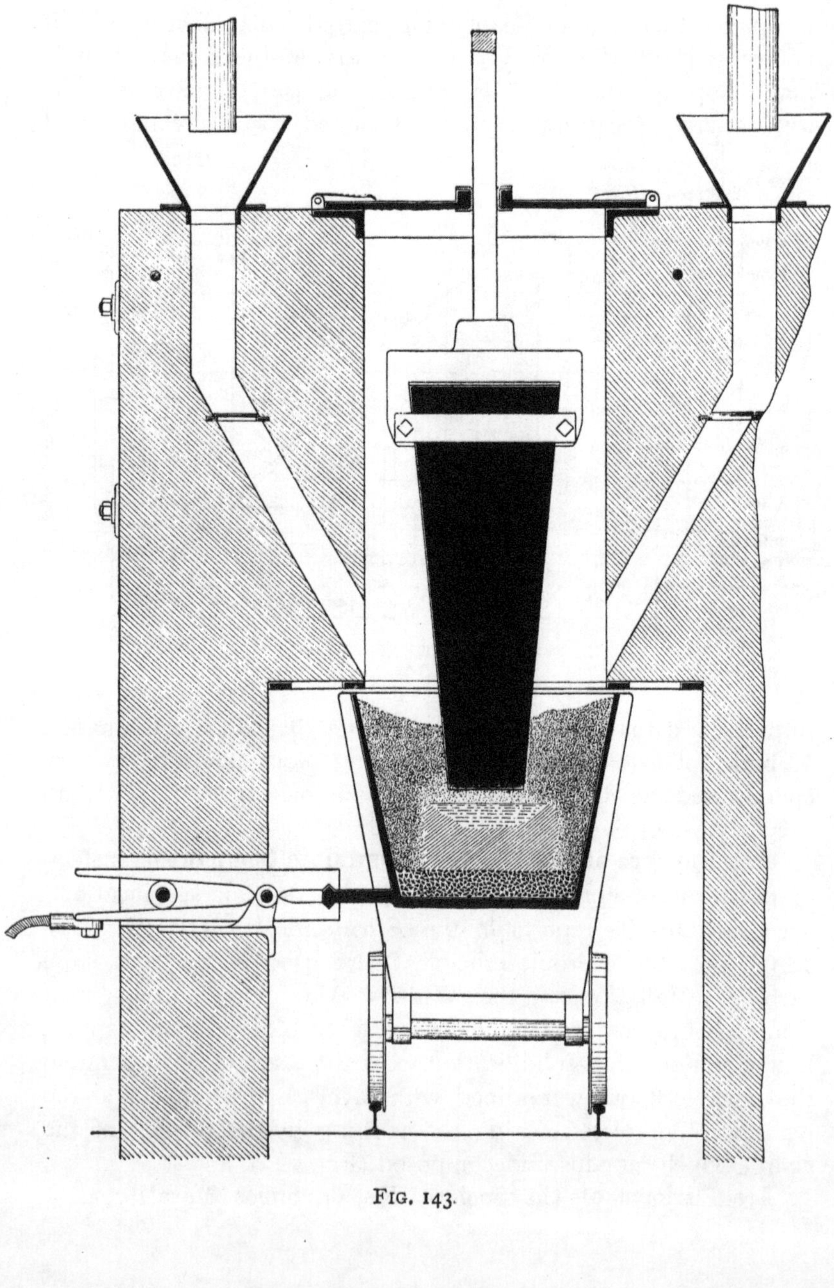

FIG. 143.

whereby the smelting was continued until the hearth became filled with calcium carbide, which, on account of its high temperature of fusion solidified to a block as fast as it was removed from the zone of the arc, also formed the first practice of the works operating according to and under Bullier (Société des Carbures Métalliques). These works, however, used pit furnaces with the

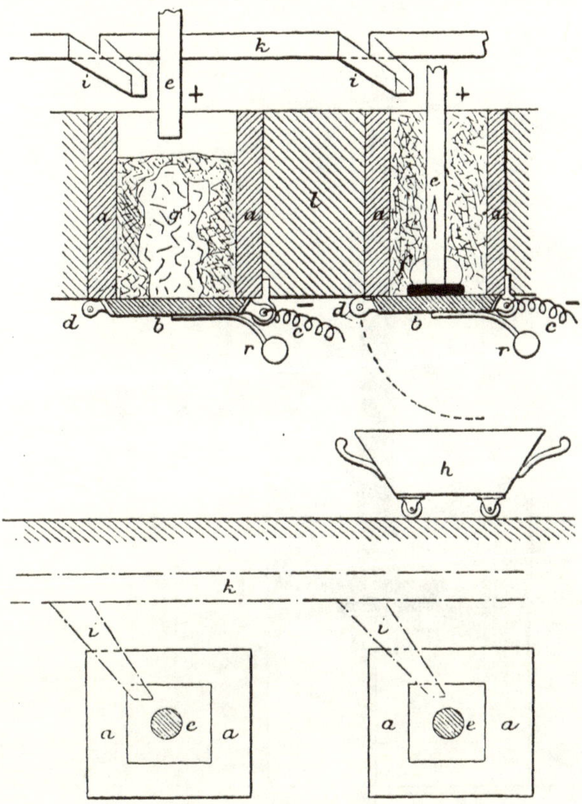

FIGS. 144, 145.

soles arranged to swing open instead of the American tip waggons. The pit furnaces consisted of carbon plates jointed into an iron frame. As in the Willson furnaces, the melting started on the furnace floor, the pits of which were constructed to heights of 1000 and even 1500 mm. (3 ft. $3\frac{3}{8}$ in. to 4 ft. 11 in.), with diameters of 600 to 1000 mm. (1 ft. $11\frac{5}{8}$ in. to 3 ft. $3\frac{3}{8}$ in.).

According to the speed with which the top electrode was raised, with naturally a constant delivery of the lime and coal mixture, blocks were obtained one-half and even two-thirds of the pit diameter up to a height of 800 and even 1000 mm. (2 ft. 7 in. to 3 ft. $3\frac{3}{8}$ in.).

Several furnaces were mostly combined in one block of

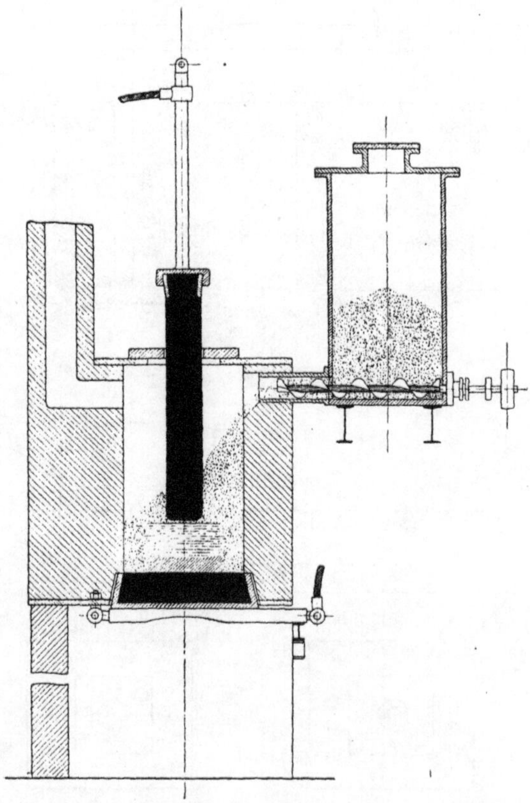

FIG. 146.—Furnace for 1500 ampères. Scale $\frac{1}{30}$.

masonry, as will be seen from a diagrammatic sketch accompanying a report by Pelissier.[1] In the furnaces described here the sole of the furnace, which is arranged so that it can be revolved on a hinge, is kept closed during the working by a balance weight and a bolt; it also forms the cathode of the furnace. At

[1] "L'Eclairage Electrique," 1896, vol. 8, p. 504.

the start of the operation the anode, also consisting of carbon, is lowered near the sole of the furnace until an arc is formed. As then a portion of the charge comprised of lime and coke becomes melted, a small hollow forms round the end of the anode, which gradually gets filled as the mass from above melts away and the anode is drawn up higher and higher, so that towards the end of the operation the furnace pit contains a block of molten material, as is shown in Fig. 144, on the left. The walls, which are made of ordinary fireproof material, remain protected by a layer of the unmolten charge. After the smelting product has cooled down, the floor plate is opened; the whole contents of the furnace are then emptied into a waggon pushed underneath, from which the block of calcium carbide is removed for packing, or for direct working to acetylene, the unreduced portion of the charge going back again into one of the furnaces.

A single furnace of this system is shown in Fig. 146.

Tenners' Arc Furnace.—Tenners' furnace[1] was also equipped with a portable hearth, which, however, in contrast to that of the Willson Aluminium Company, was flat. The furnace pit and the hearth plate were, however, so wide that outside the melting-zone a pretty thick layer of charging material still remained undecomposed, so that the furnace walls are uninfluenced by the fusion and the escaping heat (Fig. 147).

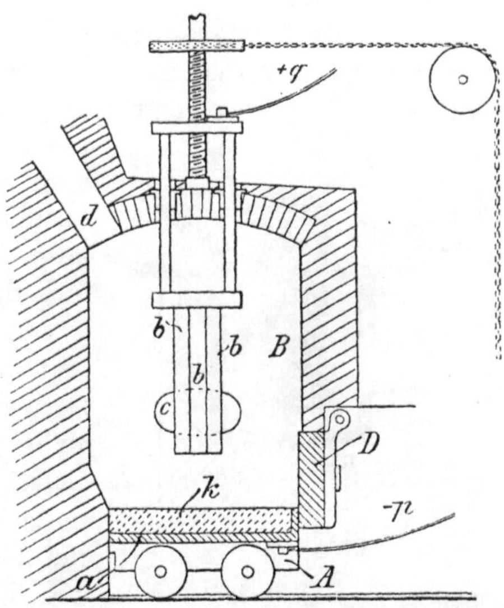

FIG. 147.

Clarke's Arc Furnace.—A pattern of simplicity of furnaces of this class is that devised by Clarke,[2] in which the working starts with

[1] German Patent, No. 88,364.

[2] United States Patent, No. 552,890.

quite a low hearth, the sole of which forms one of the electrodes (Fig. 148).

As soon as the comparatively flat layer of the first charge has been reduced, the side walls of the furnace are elevated by laying bricks on the lowest layer, fresh charge is then added, and the second electrode, which is suspended in the furnace chamber, is gradually raised. The furnace is built up during the work as the charge gets melted and reduced.

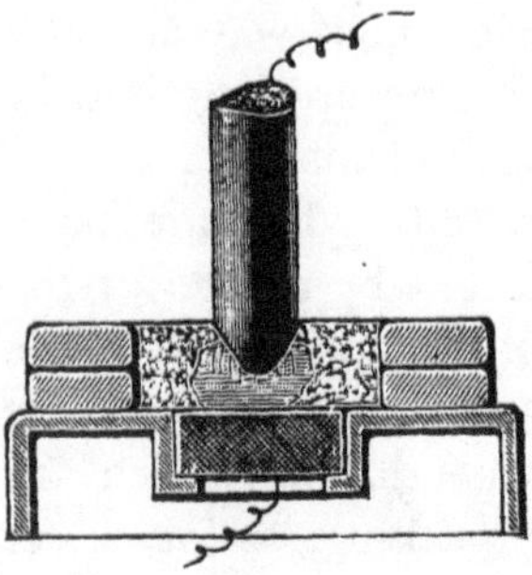

FIG. 148.

Morehead's Arc Furnace.—Morehead[1] proceeds in a similar manner to protect the electrodes from over-heating. He, however, builds up three of the side walls as high as they are intended to go; the third one, like Clarke's furnace, is only raised during the working as the furnace is filled.

Rathenau's Furnace.—In Rathenau's furnace[2] the charge is to be introduced so that the top electrode is protected by it as far as possible. To prevent, on the one hand, the removal of too much raw material on tapping the flux, and, on the other hand, the introduction after the tapping of too large a quantity of fresh material into the arc, a slide consisting of a carbon block, or some such material, is placed below the feed hopper, so that by its means the charging can be interrupted. Sufficient time before tapping the furnace the slide is closed. The raw material in the furnace is then consumed until it is time for tapping; when this operation is finished, the slide is then gradually opened so that the colder material cannot pass too suddenly into the arc and so lower the surrounding temperature that a partial solidification of the newly formed product would take place (Fig. 149).

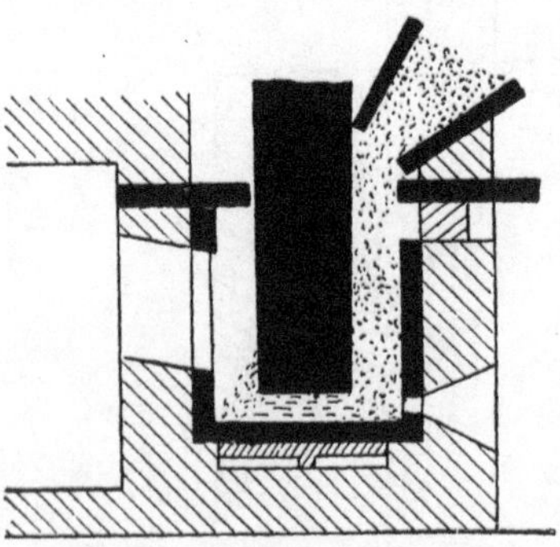

FIG. 149.

[1] United States Patent, No. 583,498. [2] German Patent, No. 99,232.

The King and Wyatt Arc Furnaces.—As simple as was the construction of the resistance furnace of King and Wyatt mentioned on p. 74, so little suitable did their arc furnaces[1] appear which were arranged to work with hollow electrodes. The top electrodes are passed vertically into a pit provided with a conducting smelting hearth, or floor; in small furnaces they are in the form of hollow tubes, and in large furnaces in the form of hollow bodies composed of carbon plates into the central portion of which the charge is distributed. The larger the circumference of such an electrode the more unequal according to actual experience is the heating, in consequence of the almost uncontrollable wandering of the arc.

The Furnace of Bresson and Pacotte.—A Siemens crucible arranged for tipping and for larger charges has been designed by Bresson and Pacotte.[2] The graphite crucible, which has an external covering of magnesia and is contained in an iron shell, is connected to the circuit at the bottom (Fig. 150).

Rotary Appliances in Carbide Manufacture. — Although block smelting in the carbide industry required a light, vertical movement of the electrodes, some designers have adopted the principle, first applied by *Kiliani*, to provide the electrodes with appliances so that they can be moved with a circular or pendulum motion over the surface of the furnace charge. Devices of this description have nowhere found lasting application; they load up the plant with machinery which is best avoided.

Another construction, due to the readiness with which the calcium carbide solidifies inside the furnace space, is that of the introduction of the movable hearth.

Naturally, the rotary aluminium furnace as supplied in the Héroult Works during the first years of working would also be suitable for operating with the arc.

The Arc Furnaces of Heath and Clarke.—Heath's[3] carbide furnace with its revolving hearth mounted on a vertical central axle presents not only nothing novel, but also nothing worth imitating in contrast with the above-cited aluminium furnace.

[1] United States Patent, Nos. 562,400, 562,404, of June 23, 1896.
[2] English Patent, No. 2790 of 1897.
[3] United States Patent, No. 586,686.

The same applies to a furnace by Clarke.[1]

The Siemens and Halske Arc Furnaces.—Five furnace constructions were brought out by Siemens and Halske in the years 1897, 1898, and 1899, partly with stationary and partly

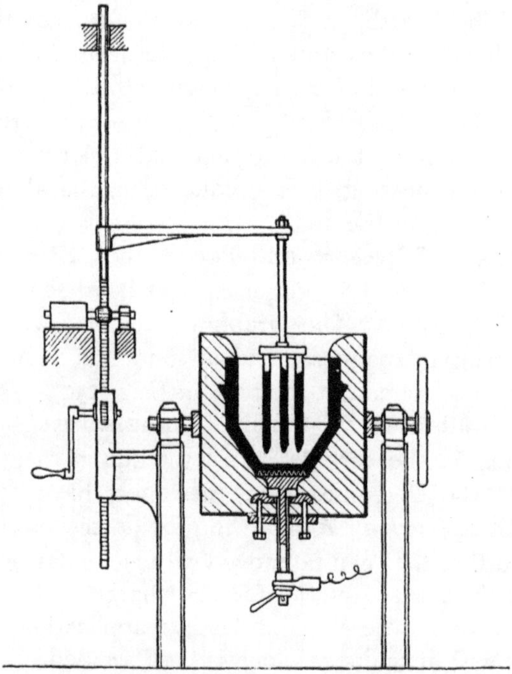

FIG. 150.

with revolving hearths, of which three were intended for the manufacture of carbide and two for the extraction of volatile metals. They have, however, not been introduced for these or for other purposes.

Bradley's Arc Furnaces.—As suitable as is the method in itself of removing easily solidified products of the smelting process by means of a wheel-shaped revolving hearth, nevertheless, the method of smelting the charge chosen by Bradley[2] in his first furnace construction, coupled with the utilization of

[1] United States Patent, No. 602,815.

[2] German Patent, No. 98,708 ; [English Patent, No. 27,888 (1897)].

the finished and mostly solidified product as one of the electrodes, appeared at the start to bring difficulties. In consequence of the large number of contacts necessary for maintaining connection with the molten product, the apparatus did not receive a construction which could be called simple. The diameter of the wheel was to be 4500 to 5000 mm. (14 ft. 9 in. to 16 ft. 5 in. app.) (Figs. 151 and 152).

The rim of the wheel is about 500 to 1000 mm. (1 ft. $7\frac{11}{16}$ in.

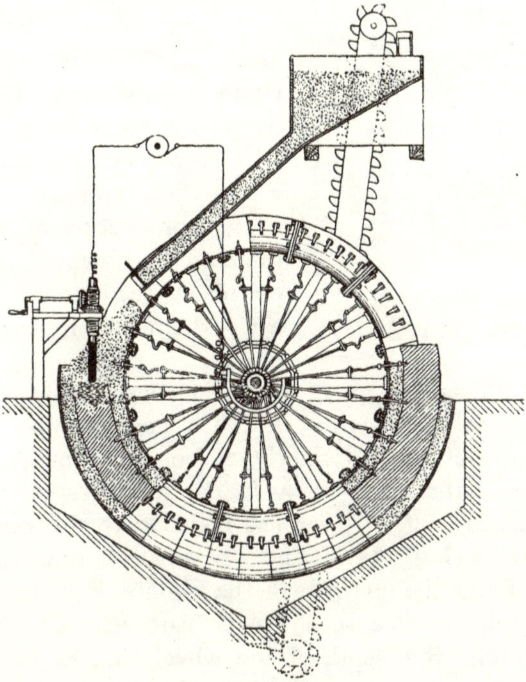

FIG. 151. Scale $\frac{1}{100}$.

to 3 ft. $3\frac{3}{8}$ in.) in width, and is recessed in the form of a semicircle; at definite distances apart it is provided with bolts revolving on a pivot, with their front ends bored out to receive the iron pins on the plates which are to be fastened to the circumference of the wheel. By means of copper pins, a commutator, and brushes, the current is conveyed from the supply to the furnace charge, while the second pole of the arc, a carbon rod, is connected to the circuit through a cap on a crane. The charge is conveyed to the furnace from the bin storing the material through a chute provided with a slider.

The wheel-pit is made with sloping walls, so that the powdered material which comes out of the wheel can gravitate to an elevator, which returns it to the bin, from which it is again discharged into the furnace.

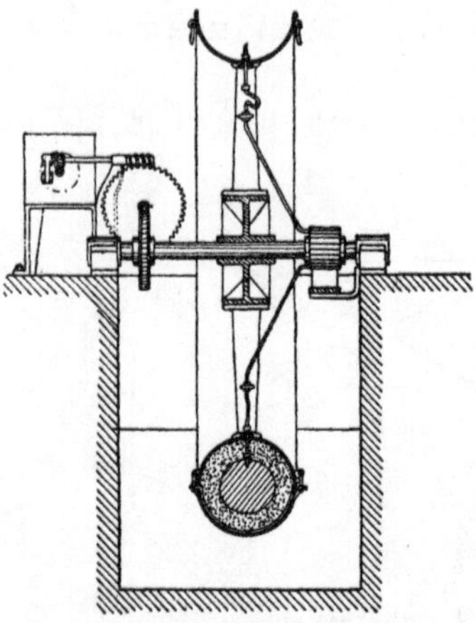

FIG. 152. Scale $\frac{1}{100}$.

The further the wheel is turned after the division of the operation, the further the individual clamping-pieces of the wheel circumference separate, and the mass gradually cools down while it is still enclosed within the clamps.

In this manner an uninterrupted production of the desired material is obtained, as the full peripheral sections are emptied by a workman and replaced by fresh ones.

In 1903 Bradley[1] made known a considerably simpler form of his furnace. The wheel rim has now a trapezoidal cross-section, but consists, as formerly, of iron with linings of graphite plates. The latter are held by fixings in corresponding grooves of the iron rim (Fig. 153), or are fastened to the rim by means of screw bolts (Fig. 154). The whole wheel rim, together with the lining, forms the one pole of the furnace, in which the charge is fused by means of a second wide carbon block which almost completely fills the depth of the wheel rim, so that the fused charge comes into direct contact with the carbon lining, *i.e.* receives contact direct from the walls of the wheel rim. The operation is conducted as in the first furnace, but when the charge falls in, iron cover plates (also furnished with graphite coverings) are put in round the periphery, to be taken off again after the completion of a rotation of about 180° for the purpose of removing the solidified product.

Bradley asserts that the fused material very rarely becomes

[1] United States Patent, No. 723,643 of March 24, 1903.

baked hard on the graphite plates. The solidified mass can, therefore, without injury to the graphite lining, be very easily broken out of the mould, which widens outwards.

The Arc Furnace of Nicolai.—Nicolai's ring hearth[1] of U-shaped cross-section moved on a horizontally-mounted wheel

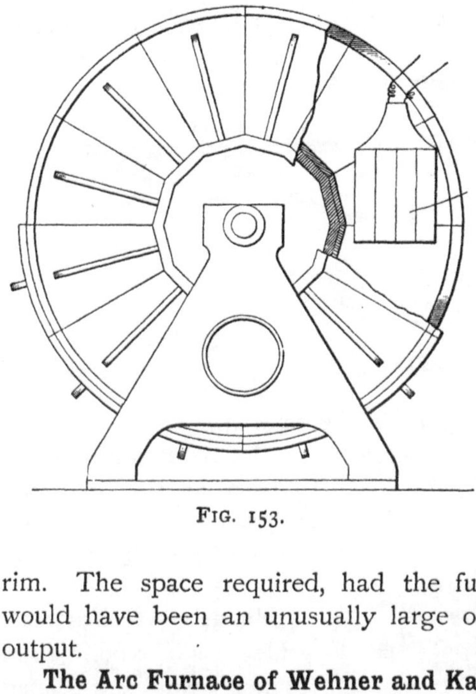

FIG. 153.

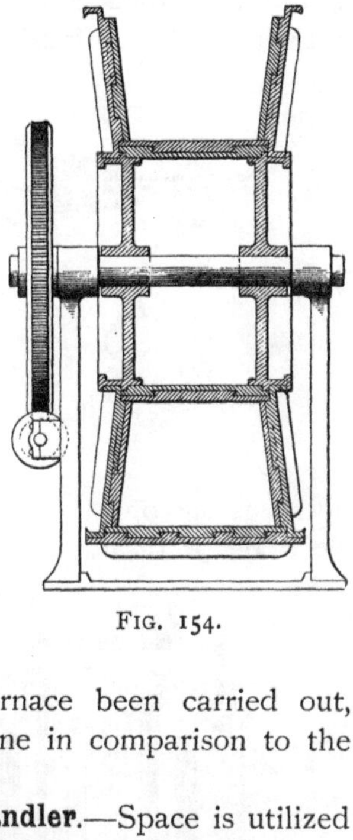

FIG. 154.

rim. The space required, had the furnace been carried out, would have been an unusually large one in comparison to the output.

The Arc Furnace of Wehner and Kandler.—Space is utilized to better advantage in the furnace[2] of Wehner and Kandler.

In judging all these furnaces, including the belt conveyor furnaces of Roberts with resistance or arc heating, it should be borne in mind that only a portion of the charge should be fused to prevent too much damage to the costly hearth. Thus, undecomposed charge and fused material leave the furnace in a very hot state, and, consequently, considerable quantities of both are able to continue burning, especially the coal still remaining in the undecomposed charge, the ash of which mixes

[1] "Jahrbuch der Elektrochemie," 1898, 5, 323.

[2] German Patent, No. 103,058 of February 19, 1898; [English Patent, No. 25,300 of November 30, 1898].

with the less changeable metal oxide. The latter can still be used, but it will be less pure.

Berthelot's Acetylene Arc Furnace.—The first of the *closed furnaces for working under pressure, in vacuo, or in a definite gas atmosphere* is the small arrangement with which Berthelot in the year 1862 combined hydrogen and carbon to acetylene in the arc. It consisted of a small glass globe with two tubular

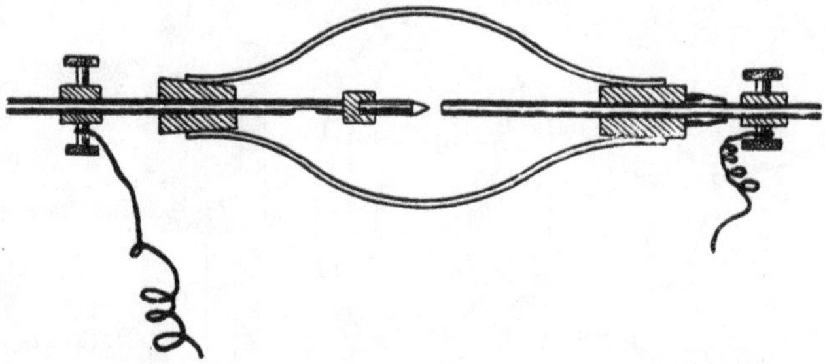

FIG. 155

openings at opposite ends of a diameter through which the electrodes were introduced in plugs. The electrodes were hollow, to pass in the hydrogen and carry off the gases produced by the reaction (Fig. 155).

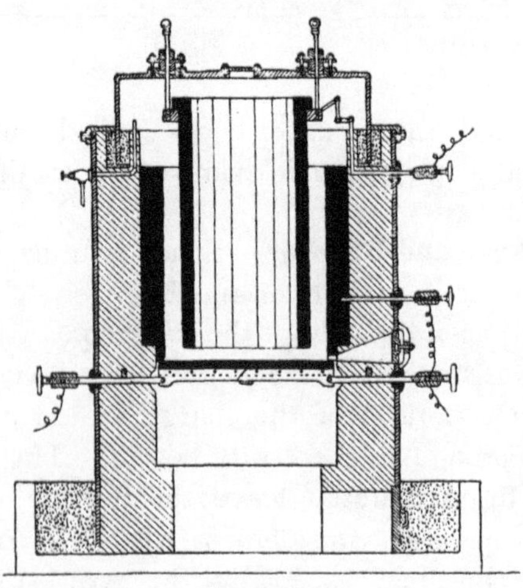

FIG. 156.

Menge's Arc Furnace.—Menge's experimental furnace[1] was constructed like an arc lamp. It was made complete for smelting experiments by arranging its lower fixed electrode in a crucible composed of conducting material. The whole was placed in a boiler furnished with steam pipes, safety valves, electrical connections, pressure gauges, and other measuring instruments.

[1] German Patent, No. 40,354 of 1886.

The Arc Furnace of Eldridge, Wright, and Clark.—Eldridge, Wright, and Clark[1] have constructed a furnace for working on a

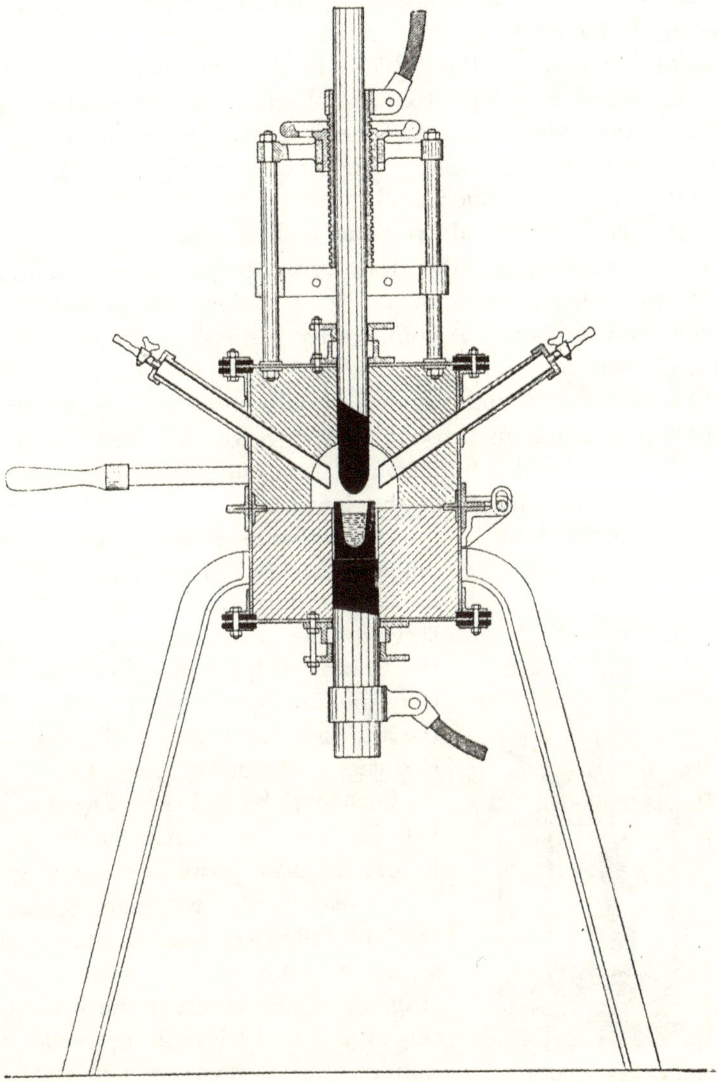

FIG. 157. Scale $\frac{1}{10}$.

larger scale. It contains a crucible inside a shell of sheet metal

[1] United States Patent, No. 583,618.

lined with masonry. The crucible consists of a carbon cylinder and a metal plate covered with carbon, and its floor forms the one electrode. The other electrode, also a carbon cylinder, hangs centrally in the crucible and receives the charge. The arrangement of the parts of the furnace is at once evident from the sectional drawing in Fig. 156. An absolute vacuum is, of course, out of the question, as here and there sand joints are made use of. The object of thus enclosing the furnace is only to exclude air and other injurious influences.

The Arc Furnace of Poulenc and Meslans.—For experimental working in my laboratory a furnace proved very suitable which was constructed with a few alterations on the design of Poulenc and Meslans. A comparatively small smelting space is formed inside an iron box, consisting of two parts, by heavily lining it with magnesite. The electrodes enter the furnace space at the top and bottom through stuffing-boxes, and are insulated from the shell of the furnace and one another. Small carbon crucibles can be placed on the lower electrode, and there are openings in the side walls which are closed on the outside by pipes with caps or rubber-tube connections attached to them. These serve for the introduction of the material to be fused, for the observation of the smelting operation, as well as for the admission and escape of the gases and vapours (Fig. 157).

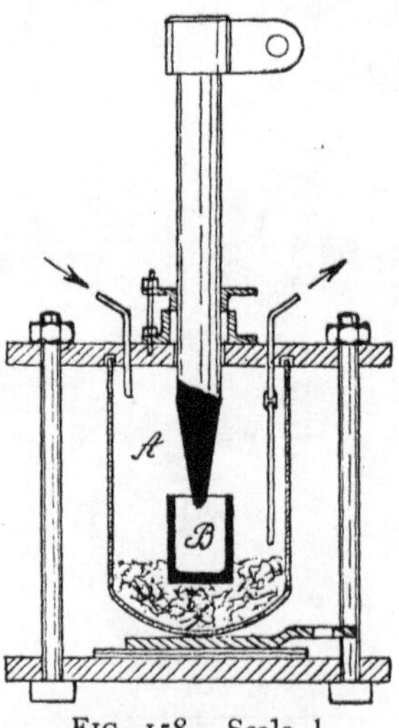

FIG. 158. Scale 1/10.

Borchers' Sealed Arc Furnace.—For carrying out experiments in a hydrogen atmosphere necessitating a closer sealing of the furnace space, I had the following small furnace[1] put together from instrument parts at hand (Fig. 158). A represents an iron vessel, the floor of which is covered with iron turnings, which serve to support the carbon smelting crucible B, and at the same time to conduct the current. A is insulated with asbestos from the top and bottom plates and made air-tight as

[1] "Metallurgie," 1904, 1, p. 413.

far as possible. A pointed electrode 40 mm. ($1\frac{1}{2}$ in. app.) is passed into the furnace space through a stuffing-box placed centrally in the upper plate and having tarred asbestos cord packing. Two pipes which are screwed in admit the hydrogen and allow of the escape of the burning gas.

Phosphorus Arc Furnaces of Readman, Parker, and the Electric Construction Corpn.—Mention should be made here of the furnaces which, according to patent specifications and other publications, are intended for the production of phosphorus, although it is improbable that they came into use as described and sketched. Thus, according to the first patents of Readman and Parker and the Electric Construction Corporation in the years 1888 to 1890, furnaces as represented in Fig. 159 were used. They were pit furnaces provided in the lower portion of the pit with carbon rods to start the heating, whereas higher up in the hearth were the heavier arc poles. A hopper with two slides and a spiral conveyor gear is provided for dealing with the charge; in addition there is an opening for leading off the gases produced by the reaction and the phosphorus vapours, and a tap-hole is also provided for the molten slag.[1]

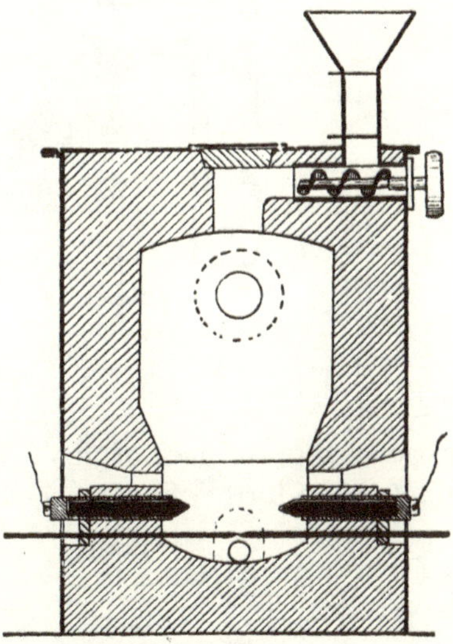

FIG. 159.

In later patent specifications resistance heating by radiation is provided for, resistances being arranged exactly as in Borchers' experimental furnace,[2] but above the charge.

The Liebmann Distillation Arc Furnace of the Deutsche Gold- und Silber-Scheideanstalt.—The furnace of the German Gold and Silver Refining Institution, provided with distillation attachments at the instigation of Liebmann, is an exceedingly efficient furnace, especially for experiments on the smelting of

[1] German Patent, Nos. 55,700, 107,736, 112,832.

[2] Cp. p. 62, Fig. 80.

charges which contain volatile elements (phosphorus) and compounds. After the explanations given on p. 98 of the first

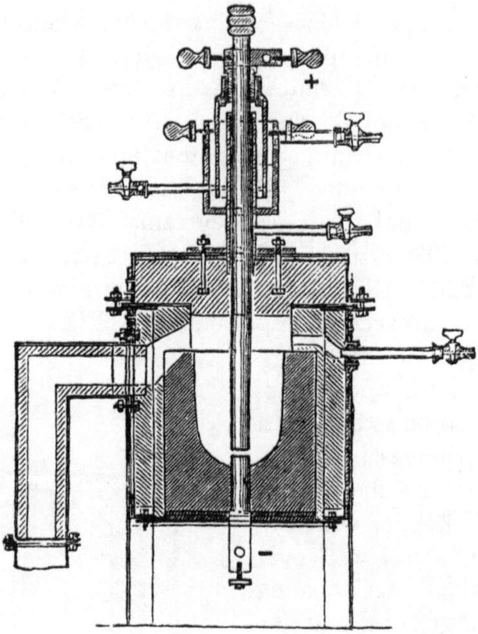

Fig. 160.

experimental furnace of this firm, the sketch depicted in Fig. 160 requires no further explanation.

CHAPTER V

COMBINED RESISTANCE AND DIRECT ARC HEATING

Introduction.—Strictly speaking, all electric furnaces with direct arc heating are no longer pure arc furnaces from the moment the molten material which has accumulated in a more or less thick layer on the floor of the furnace, acts as one of the poles of the arc, for the layer of fused material often forms a by no means inconsiderable heating resistance. In many experiments with arc heating the dimensions of the electrodes are intentionally so chosen that they themselves, at any rate at their ends, would become strongly heated in the furnace by the current, even though it would not result in any arc heating at all. In experiments on the small scale one is compelled, for the attainment of the highest degree of heat, to make the electrodes too thin for the current, if only at their ends, as they would conduct away too much heat to the outside on account of their good heat conductivity, unless strongly heated by the current which they are carrying.

The Resistance-arc Furnaces of L. Clerc.—As the first furnaces of this kind I should like to denote those of L. Clerc of the years 1880 and 1881. We recognize to-day, at the first glance, that they could only work if the furnace walls, or the charge, or both, get connected to the circuit as intermediate electrodes and, consequently, also as heating resistances.

In the year 1880[1] Clerc patented an electric lamp, which, under the name "Lampe Soleil," was shown at the International Electrical Exhibition of 1881. In this lamp the electrodes were introduced through cavities in a limestone or magnesia block, inclined at an acute angle downwards, in such a manner that

[1] French Patent, No. 134,519 of January 13, 1880.

underneath a perfectly free arc ensued. After the arc had started, the carbons could be drawn back sufficiently for the layer of lime or magnesia lying between the ends of the carbons to take part in the conduction of the current, producing a long, unusually brilliant arc. The lamp had no success. The lime became fused and dripped down; even the magnesia was decomposed and volatilized. But the unfavourable experiences with this arrangement in its capacity as a lamp led Clerc to use it as a furnace for high temperatures, which he so arranged that it could be worked with an explosive mixture of oxygen and hydrogen as well as with electricity.

In fact, Clerc at first utilized his lamp itself as the source of heat by placing it on a block of limestone as the hearth, in the hope of being able to fuse sufficient lime in this way to cast from it the guide-blocks for his arc lamp carbons. After several experiments, he finally embodied the views arising from his experiments on the arrangements of furnaces for very high temperatures in the description of a Belgian patent taken out on the 9th of July, 1881.

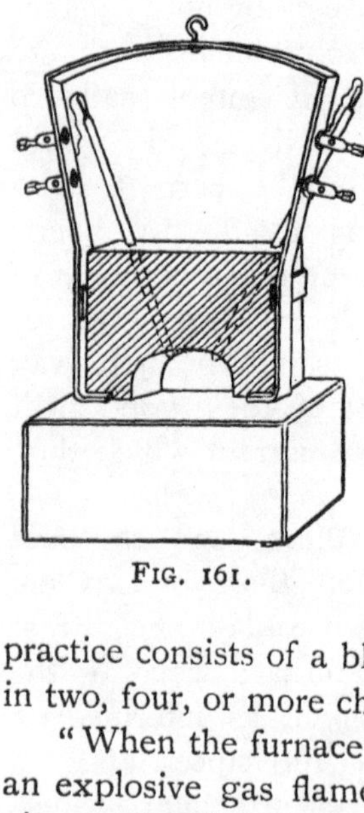

Fig. 161.

"Our invention is for the purpose of producing and usefully applying the enormous quantities of heat which an explosive gas flame or an electric arc can produce alone, or together, when they are in constant contact with the same surfaces of the substance to be heated. The apparatus or furnace with which we carry out this invention in practice consists of a block of resistance material in which we let in two, four, or more channels.

"When the furnace is to be brought to a high temperature by an explosive gas flame, each pair of channels receives two gas pipes to admit the hydrogen and oxygen in the known proportion.

"When the apparatus is to be heated with an electric arc we provide these channels with carbon rods, which we connect to the circuit of a dynamo machine.

"These rods are moved by counterweights or springs. Their ends are let in the solid mass, and the arc passes through the two orifices of the channels which enclose them. As regards the quality of the block, it suffices for it to be composed of a resistance mass; for this purpose we use, if not exclusively, nevertheless preferably, a mixture of calcined magnesia, and a metal oxide which is compressed under high pressure. The internal form of the electric furnace depends on the substance to be subjected to the high temperature. As a rule it consists of a simple, circular cavity on which a movable cover can be placed.

"In certain cases this cover can be solid, or is perforated for the introduction of the materials to be worked.

"It is, therefore, readily seen that with the arrangement indicated the mass to be handled is always in contact with the explosive gas flame or the arc, the position of which remains uniformly unchangeable.

"The applications of this furnace are numerous.

"The arrangement with which an arc is used, and with which, for instance, carbon can be converted into graphite, can be applied to many purposes.

"The adjoining figures 161 to 164 are a true representation of our construction.

"A is a block composed of a solid mass; it can be made of one or several pieces sheathed with iron rings.

"C is an open space; its dimensions differ according to the use to which it is put.

"D is a lid which keeps the heat in the furnace in case it is desired to attain higher degrees of heat. It is furnished with one or several openings, O, for the admission of material to be worked up and for the escape of gases or vapours.

"When the furnace is worked electrically, the smelting takes place within one or several arcs formed between equal pairs of carbons B B, B′ B′, B″ B″, etc., the ends of which are let in the block and pass through the orifices O, O′. These carbons may be inclined (Fig. 161), horizontal (Figs. 162, 163), and vertical (Fig. 164).

"The cross-section of these carbons is circular, but other forms may also be advantageous: angular, hollow, and cylindrical, etc.

"The cross-section may be very large; its mass can be mixed with materials to increase its durability or raise its capacity as a conductor, or its interior can be filled with such substances.

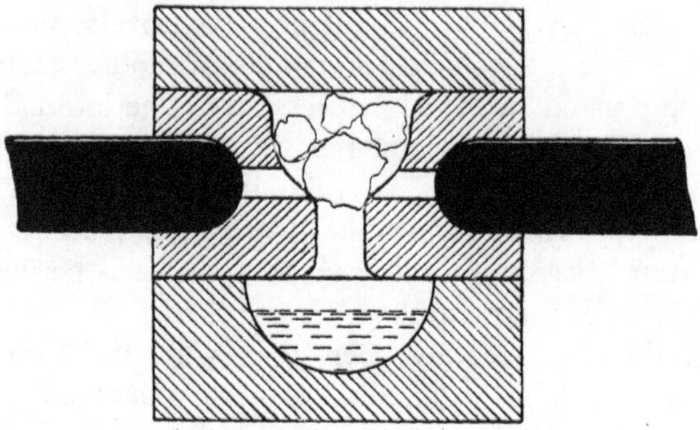

FIG. 162.

"The hollow pencils can serve the purpose of conducting the streams of explosive gas into the furnace.

"The feeding-in of these carbon rods can be done by a spring

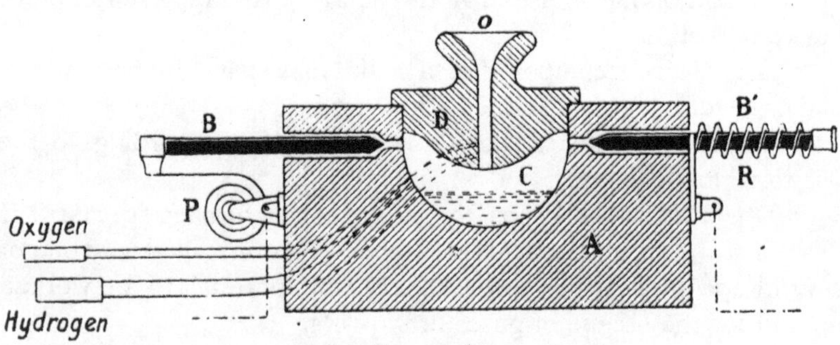

FIG. 163.

mechanism, either in the shape of spiral springs, R, R′, or simple metal strips. These springs can also be replaced by weights.

"The current may be produced by one or more dynamos or other electrical source.

"If the furnace works with the explosive gas flame, hydrogen and oxygen are led into the interior of the furnace through one or more pairs of tubes; these convey the gas from the gasholders in the fixed proportions for the attainment of the greatest degree of heat.

"However great the number of pairs of carbons or gas tubes may be, they must in either case be symmetrically arranged with

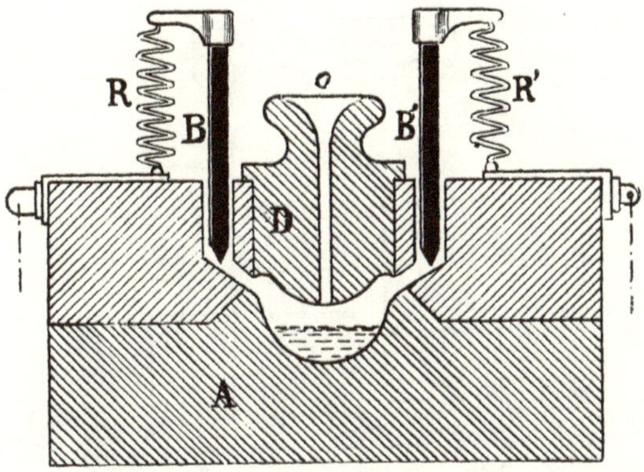

FIG. 164.

reference to the centre of the furnace, in order to ensure the regular and rapid heating of the material to be worked up."

The Resistance-arc Furnace of the Société des Carbures Métalliques.—Not till very much later did the Société des Carbures Métalliques construct a combined resistance and arc heating furnace for block-smelting as developed in the manufacture of carbide, with the avowed intention of connecting the product of the smelting between the poles of an arc as the intermediate electrode and heating resistance (Fig. 165).

Now, although the hot calcium carbide is a fairly good conductor, the increase in height of the carbide block on the melting hearth, *i.e.* on the original electrode floor of the furnace, means a lengthened path for the current and, consequently, an increase in the resistance. The Société des Carbures Métalliques,[1] therefore, made a change by arranging both electrodes on the surface of the fused product, but dipping into the charge, so that the current had to pass from the one electrode to the fused material, and from this to the other electrode. It is true that in this manner two arcs could arise, if the current, as is mostly the case in carbide manufacture, did not find a path of less

[1] French Patent, No. 261,337 of November 16, 1896.

resistance in the hot layer of the charge floating on the molten liquid than is formed by the layer of gas between the electrodes

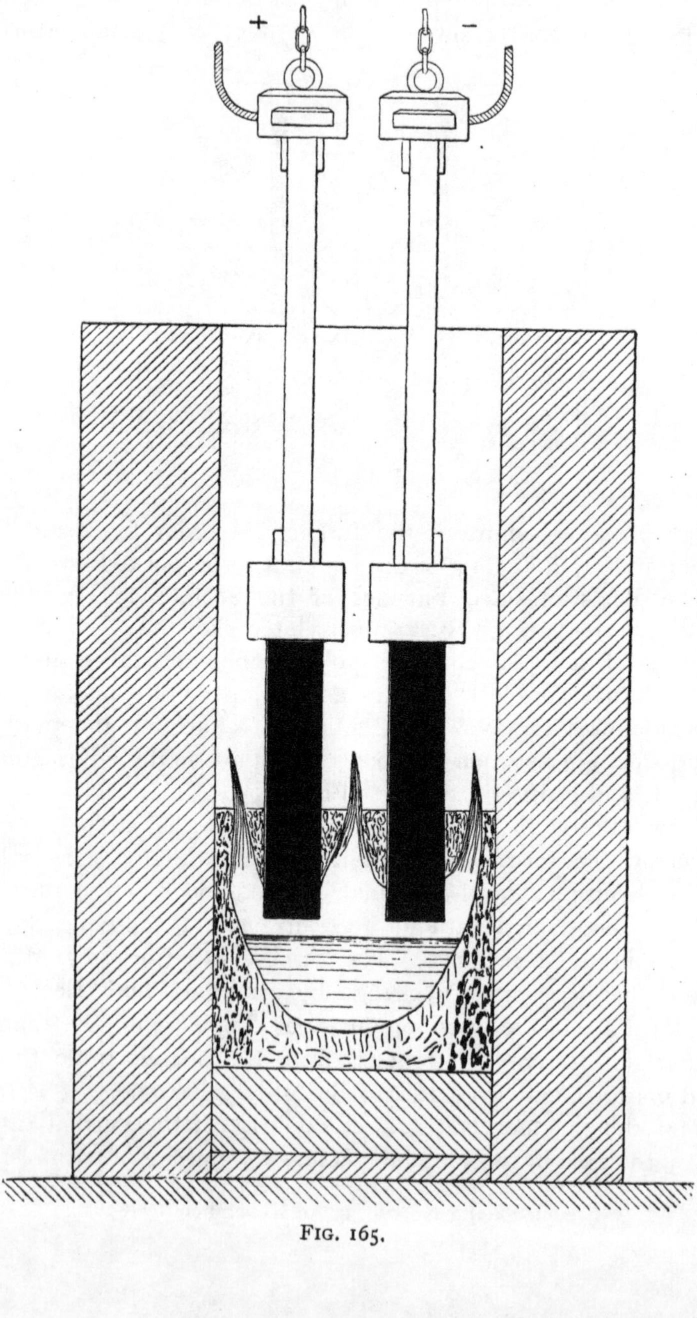

FIG. 165.

and the fused mass, but we know that the production of an arc in an atmosphere of lime vapour only requires a comparatively small E.M.F. It was, further, important for working to be able to keep the resistance uniform until the furnace pit was full, instead, as formerly, of having to work from the start of the operation to the finish with the carbide block, with a resistance which increases in a by no means regular manner. Cracks and accumulations of lime, etc., in the cooling carbide block could often, very perceptibly, increase the resistance in the circuit. The source of trouble remained unknown until the operation was stopped, and, therefore, could not be removed during the working. In the new furnace the total path of the current was transferred to the surface of the working hearth; it thus always remained accessible. Only the molten portion of the product of the furnace took part in conducting the current, and acted as the heating resistance, producing heat in a manner favourable to the operation, and imparting heat to the layer of the charge floating on it. We have here one of the realizations of the methods of electrical heating already given by Clerc in 1881.

Horry's Drum Furnace.—Horry has very ingeniously avoided the drawbacks which occur in block-smelting without, however, losing the heat which is used in other methods of working to keep the finished carbide in the liquid state. In the description of his furnace he rightly points out that in block-smelting as hitherto practised, as well as in the process of tapping, large quantities of carbide have to be kept warm until the furnace is full, whereby heat losses, injury to the apparatus, and other evils are difficult to avoid. Horry's furnace enables the product of smelting to be continuously removed from the region of the stationary arc, and to be finally removed from the furnace at some other part after it has cooled down.

The material of the charge is delivered between pairs of electrodes situated in a funnel composed of fire-bricks and fire-clay. In this manner it is forced to traverse the zone of the arc. A drum, **U**-shaped internally, moves round this arrangement for charging and smelting. The requisite rotation of the drum is effected through worm gearing, and its periphery can be partially covered with plates held by easily detachable bolts (Figs. 166, 167). If a diminution of the resistance between the electrodes should be indicated by a galvanometer placed in the circuit, *i.e.* an accumulation of conducting, fused products,

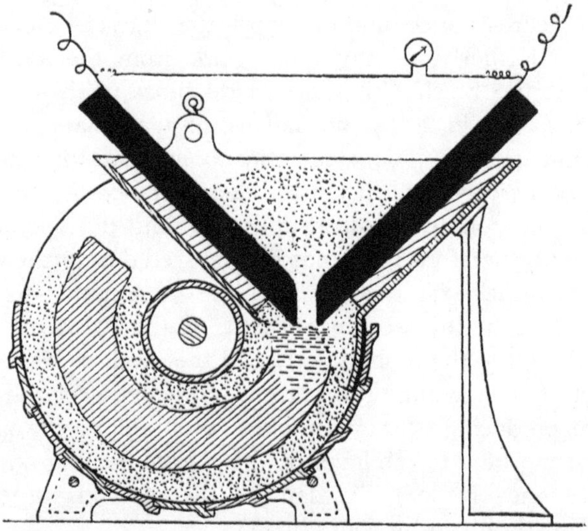

FIG. 166.

the fused material is rotated in the direction of the arrow (Fig. 166) until the prescribed strength of current is flowing through the furnace. A fresh plate is then put on, and a similar one removed at the other end of the charge. The products of the fusion thus form a cooling and solidifying ring during the rotation of the drum, at the open side of which the solidified portion is broken off (Fig. 166).

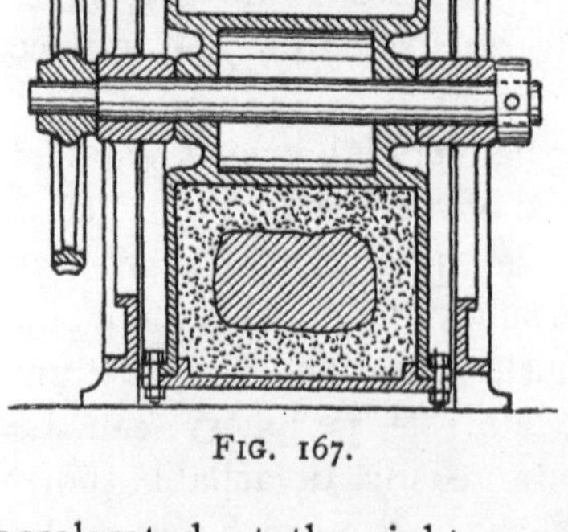

FIG. 167.

Patin's Resistance-arc Furnace.—In Patin's pit furnace the hearth is arranged so that it can be lowered, and blocks can be produced without materially stopping the working. At a definite depth the downward motion is accelerated, at the right moment a slide is pushed beneath the electrodes and above the finished block to prevent any sinking of undecomposed charge, and the block is then drawn out through a side door, upon which the hearth is again pushed up and the slider drawn back (Fig. 168).

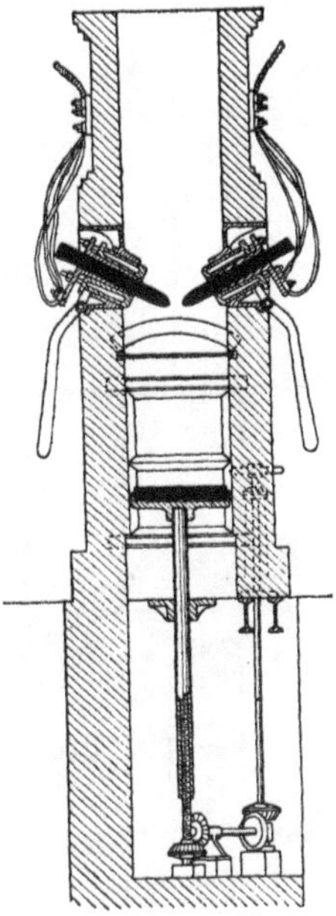

FIG. 168.

Memmo's Furnace.—Memmo also works with the same kind of dropping hearth, but surrounds the lowering pit with a hollow metal shell, in which cold air ascends for cooling the fused products. In addition, devices are provided above the smelting space for preliminarily heating the charge and for further utilizing the waste gases.[1]

Price's Resistance-arc Furnace.—In a furnace by Price an iron bed-plate covered with carbon constitutes the hearth of the furnace.[2] On three sides the hearth is closed with carbon walls; on the fourth side, towards which it slopes slightly, an opening is provided through which the products of the smelting flow into a moulding-pan firmly pressed against the floor-plate by means of a wedge. The whole of the furnace chamber is enclosed by fire-proof masonry. At the top the furnace is closed by the feed-hopper and the charge. The electrodes are fastened by means of clamps to rods provided with screw-threads. The rods pass through stuffing-boxes screwed to a supporting beam, and can be raised and lowered by screw-threaded hand-wheels. The electrodes are connected to the source of supply by means of caps. To ensure a uniform descent of the charge, each electrode is surrounded by special fill-boxes attached to the bottom of the common feed-hopper, and consisting of hollow metal bodies through the hollow space in which a cooling agent circulates by means of tubes. These hollow bodies also constitute the outlet pipes for the gases

[1] English Patent, No. 14,022 of 1897.

[2] German Patent, No. 93,798 of December 2, 1896; United States Patent, No. 572,312 of December 1, 1897; [English Patent, No. 27,301 of December 1, 1896].

produced by the reaction, which can, therefore, collect in the channel from all parts of the furnace, to finally escape from these through the flue to the chimney (Figs. 169 and 170).

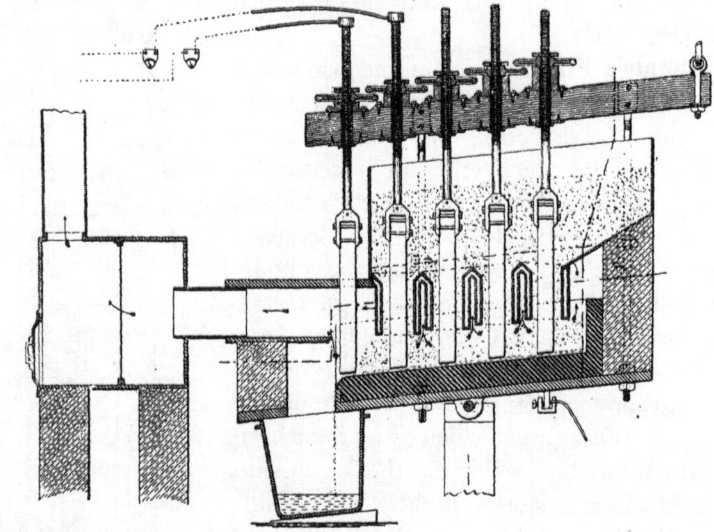

FIG. 169.

The whole furnace is supported partially by pillars and partially by beams. For this purpose the bearings are supported on the pillars, and a bolt is passed through them and two hinges cast on the bed-plate of the hearth. By means of four other hinges and bolts the plate of the hearth and, consequently, the furnace itself is hung to the beams, so that the furnace can be inclined to any extent by means of the anchored bolts.

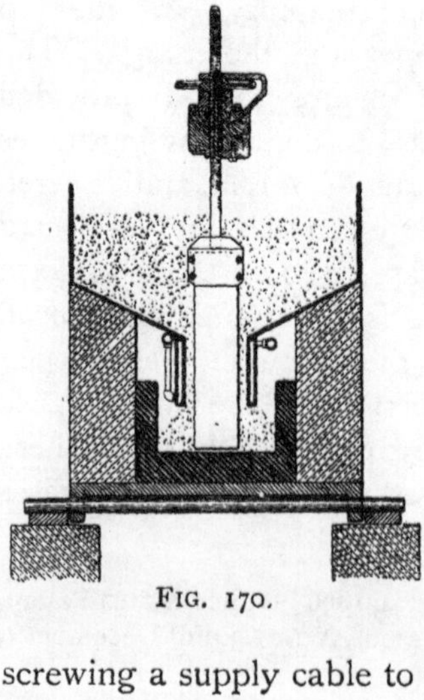

FIG. 170.

The heating is accomplished by the arc which is allowed to form between the electrodes on the one side and the plate of the hearth on the other. The hearth-plate is connected to the current source by screwing a supply cable to the iron bed-plate.

As proved by his patent specification, Price did not intend to construct a furnace with combined resistance and arc heating; these actions, however, without doubt occur from the method used in charging as illustrated in the drawing, if four of the electrodes stand completely in the charge and the fifth stands in front of the extreme slope of the latter at a distance requisite to the formation of an arc.

Borchers' Resistance-arc Furnace.—A clearer idea of a

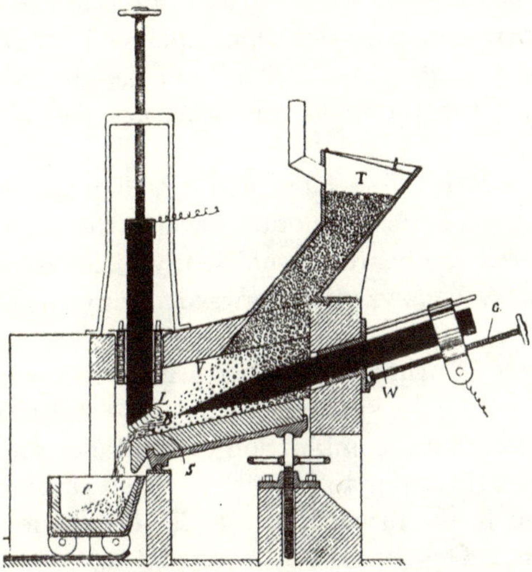

FIG. 171.

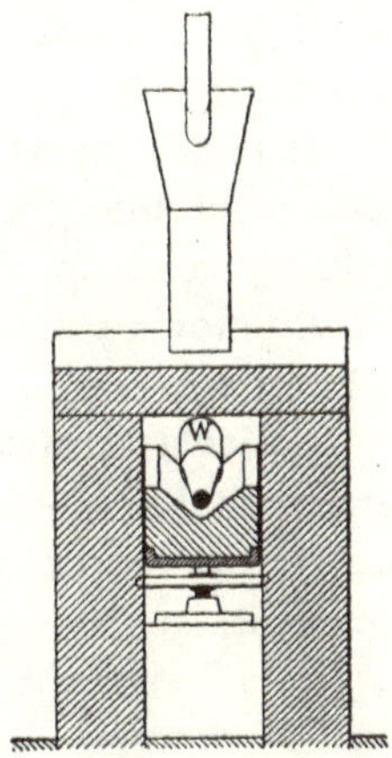

FIG. 172.

resistance and arc furnace will be gathered from the following description and illustration of a furnace expressly designed by Borchers[1] for this purpose (Figs. 171 and 172):—

The furnace consists of the preliminary heating space V and the actual smelting space S. In the upper part is arranged the feed hopper T through which the material to be acted upon slides down to the melting zone in such a manner that the charge decreases in cross-section towards the arc. Of the two carbon electrodes L and W the latter can be pointed for the purpose of increasing the electrical resistance. The molten matter flows down into the truck C, or into a mould.

[1] German Patent, No. 115,742 of June 2, 1898; English Patent, No. 6061 of 1899; United States Patent, No. 660,034 of October 14, 1900.

In starting the furnace the carbon electrode W, which later acts as the heating resistance, is brought near the end of the electrode L until an arc is formed. As the electrode W is being drawn back, with a charge which is conducting, or which becomes conducting in the heat, the part of the charge B situated between W and L gets connected to the circuit at its extreme end as an opposite pole to L. As, however, both the long-pointed electrode W and the charge B, as far as the latter is conductive, or becomes so by the heating, steadily decrease in their cross-sections towards the arc, the resistance grows in this direction, and with it the temperature of the material, so that in the vicinity of the arc only a slight increase in the temperature is still necessary to melt the mass.

If it is a question of a non-fusible substance, *e.g.* charcoal for conversion into graphite, the advancing motion towards the arc is suitably accomplished when the hearth is sufficiently inclined by slowly moving the carbon W backwards or forwards by means of a handle, G, or an eccentric.

Keller's Resistance-arc Furnace.—In his studies on the most suitable method of heating for smelting carbide, Keller[1] has also come to the conclusion that a combination of resistance and arc heating should be used in order to keep the molten carbide liquid as long as it remains in the furnace, with as little expenditure of power as possible. This furnace is equipped with two electrodes, A, B, arranged vertically. They are held in the caps C, and can be regulated as required by means of the rope or chain pulleys DG, HI. On the floor of the hearth of the furnace is a layer, K, of conducting pieces of carbon. In setting the furnace to work, both electrodes are lowered on to the carbon layer K, so that it forms the heating resistance between the electrodes. As soon as this layer has been brought to a sufficient glow, the charging is started.

It is important for the working that the resistance of the carbon layer is high enough to prevent the whole dynamo current from passing through at the start. As the fused material collects on the carbon layer, the latter being also conductive, the resistance of the furnace continually decreases, and more and more current can then be gradually passed through it. In

[1] "Jahrbuch der Elektrochemie," 7, 523 (1900).

proportion to the accumulation of the fused material in the apparatus the position of the electrodes must be altered for the maintenance of a moderate voltage and current. As will be seen from the illustration, the electrode B can only be raised and lowered, whereas the electrode A can be raised, lowered, and moved sideways; at the end of the process the latter, as shown in Fig. 173, will be over the tap-hole J.

With this disposition of the electrodes, the product of the

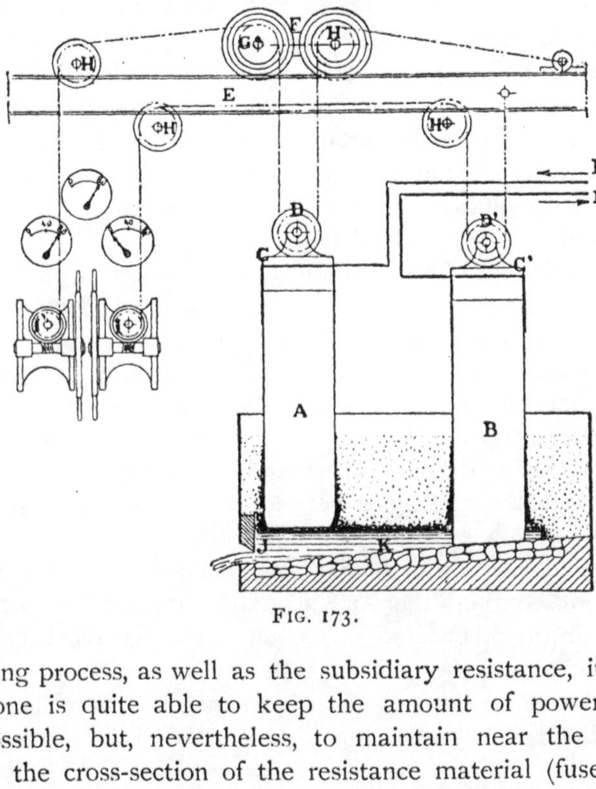

FIG. 173.

smelting process, as well as the subsidiary resistance, it is clear that one is quite able to keep the amount of power as low as possible, but, nevertheless, to maintain near the tap-hole, where the cross-section of the resistance material (fused matter and carbon floor) is greatest and consequently the actual resistance heating at this place would be least, the necessary fluid condition of the fused material by the production of an arc between the latter and the electrode A.

The Héroult Iron Furnace.—The step in the development of this heating system towards the perfection and simplicity necessary for its further application was still wanting until, in the year 1899, it was taken by Héroult[1] (the patent applications commenced at the beginning of 1900). In the same way as by his simple furnace construction and clever application of the electrolyte as the heating resistance he laid the foundations of the aluminium industry, he also invented a process, no less remarkable for its simplicity, with which he in fact accomplished the first practical success in the electro-metallurgy of iron, and in this respect need only acknowledge the competition of the Ferranti-Kjellin process. Héroult only indirectly connected the metal to be fused to the circuit ; he placed layers of a slag, *i.e.* of an electrolyte, between the metal and the electrodes as the actual heating resistance. He thus prevented any direct contact of the metal with the unavoidable carbon electrodes ; and in this manner, but only in this manner, is it possible *to refine iron to any degree of purity*, and to introduce into it any desired ingredient in the exact quantity (Figs. 174 and 175).

The metal bath of the furnace is either molten in the electric furnace itself or in some other manner, and is contained in a vessel lined with a non-conducting material, free from carbon, and of constant heat capacity. On this bath is kept a layer of slag, or other substance, which is only moderately conductive at higher temperatures. Into this the electrodes are allowed to dip, at a sufficient distance apart and sufficiently deep that, on the one hand, the resistance between the electrodes inside the slag layer is high enough to compel the current to go from the one electrode through the slag beneath it to the metal, and from this through the same slag layer to the other electrode, and that, on the other hand, no contact can take place between the electrodes and the metal. Whether an arc be maintained between the electrodes and the slag or not, the most important thing is the regulation of the distance between the electrodes

[1] French Patents, No. 298,656 of March 27, 1900, No. 305,373 of November 12, 1900, No. 307,739 of February 1, 1901, No. 318,638 of February 12, 1902, No. 320,682 of April 25, 1902, No. 328,350 of January 7, 1903, No. 336,705 of November 2, 1903 ; German Patents, No. 139,904 of July 4, 1900, No. 148,706 of July 28, 1901, No. 142,830 of February 22, 1902 ; English Patents, No. 16,293 (1900), No. 14,486 (1901), No. 14,576 (1901), No. 14,643 (1901), No. 3912 (1902), No. 6950 (1902), No. 7072 (1903) ; United States Patents, No. 721,703 of March 3, 1903, No. 707,776 of August 26, 1902, No. 733,040 of July 7, 1903.

and the metal bath, in such a manner that the layers of slag which exist between the electrodes and the metal bath remain hotter, and therefore more conductive, during the whole working

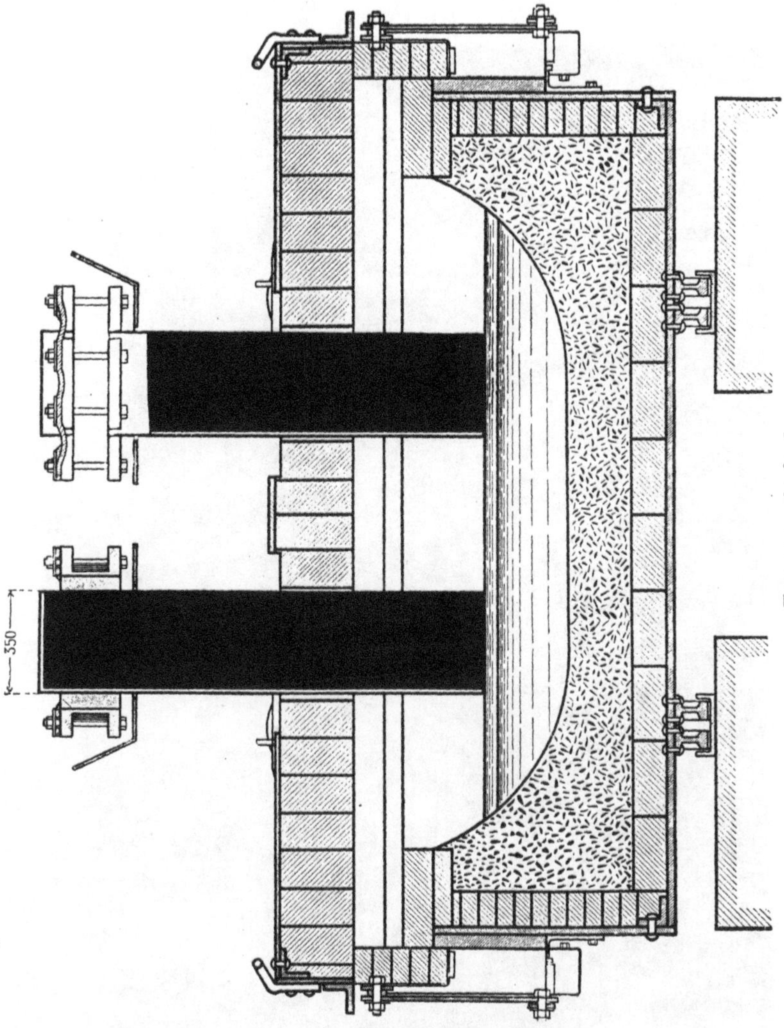

Fig. 174. Scale $\frac{1}{25}$.

process than the layer of slag between the electrodes themselves; for only in this way will the current path be as described above. It is only necessary to mention that one can act on any metal molten in this manner by means of reagents, which are placed on or introduced into the metal, whether it is to remove the undesirable constituents from an impure metal, or to introduce into it the ingredients desired.

The Héroult process can be carried out on a comparatively

Fig. 175.

small scale; for small demonstrational purposes in the Aachen Institute of Mines and Electro-Metallurgy, a plant equivalent to about 50 E.H.P. was found sufficient.

Starting with cold metal, with a 300-H.P. furnace Héroult has fused 20 kg. (44·1 lb.) of steel with one electrical horse-power in twenty-four hours; it is, of course, clear that the furnace output would be very considerably increased by charging it either completely or partially with raw iron in the liquid state.

CHAPTER VI

INDIRECT ARC HEATING

FURNACES IN WHICH THE SUBSTANCE TO BE HEATED IS IN A SPACE HEATED BY INDEPENDENT ARCS: ELECTRIC RADIATION FURNACES

Introduction.—We now turn to the most disadvantageous method of electric heating, the application of which is only justified when very high temperatures are necessary for a substance which, even under these circumstances, is non-conductive, or should in no way come into direct or indirect contact with the substance of the electrode, especially carbon, if only through the agency of the arc zone. A glance at the literature of the subject will show us, however, that this heating method has in numerous cases been proposed without reason where any other heating system would have been preferable. We know that in the arc between carbons (and these are almost exclusively used as the electrodes in indirect arc heating) temperatures of 3500° C. occur, and, in fact, there are processes and designs of apparatus for reducing conducting substances by means of this heat source which only require 1000 to 1200° C., when they could be easily and quickly reduced by heating with 1500° C.

The first furnaces of this class owe their existence to the desire to electrically melt iron ores or metallic iron.

Pichon's Indirect Arc Furnace.—In the year 1853, a French patent was taken out by the chemist, Pichon,[1] demonstrator at the École de Chimie Pratique in Paris, and an English patent, under the name of Johnson (patent agent?), for *an application of*

[1] French Patent, No. 15,880 of March 16, 1853; English Patent, No. 700 (1853).

the electric light for metallurgical purposes, and especially for the metallurgy of iron, for smelting and reducing all sorts of ores and metals without any kind of fuel. The ore to be reduced, or the metal, was to drop down between large pairs of electrodes, which are connected to the electrical circuit, and was to be melted in its passage through the arcs. In a collecting chamber, which could be heated, and was arranged below the poles, the molten masses would have time to separate out according to their specific weights.

Pichon already intended the construction of these furnaces on a fairly large scale, for he proposed to use for the electrodes rods having a cross-section of 60 sq. cm. (9·3 sq. in.), and a length of 3 metres (3·3 ft. nearly) (Fig. 176).

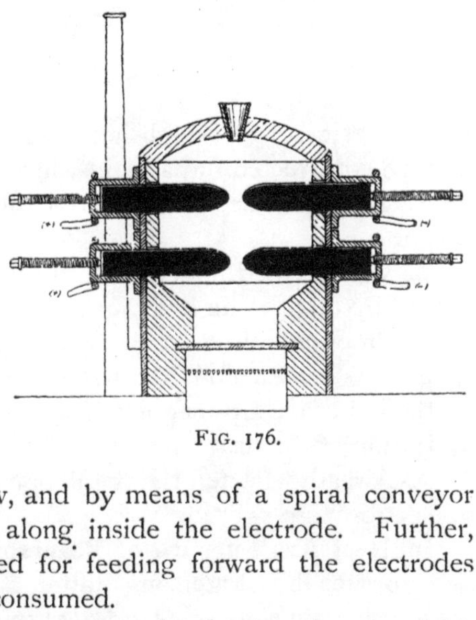

FIG. 176.

A modification of this smelting arrangement is given in the same patent specification, and consisted in the two pairs of electrodes being obliquely arranged opposite one another. The charge was to be introduced into the arc through the highest electrode, which was hollow, and by means of a spiral conveyor the ore was to be pushed along inside the electrode. Further, screw devices were provided for feeding forward the electrodes as they become gradually consumed.

The inventor, as will be seen, was far, far in advance of his day, for in an age of galvanic elements his daring plans were unfortunately unable to be realized.

The Indirect Arc Furnace of Ch. W. Siemens.—It was not until twenty-five years later that a new furnace (Fig. 177) by Ch. W. Siemens[1] became known. It consisted of a melting crucible of non-conducting material with lateral regulation of the electrodes. The one electrode was to be made hollow, so that by the introduction of gases a reducing, or neutral, atmosphere could

[1] English Patent, No. 2408 (1878).

be obtained. Hollow metal tubes arranged for water cooling were also proposed for the electrodes. With reference to what has already been said about the Siemens direct arc furnace on

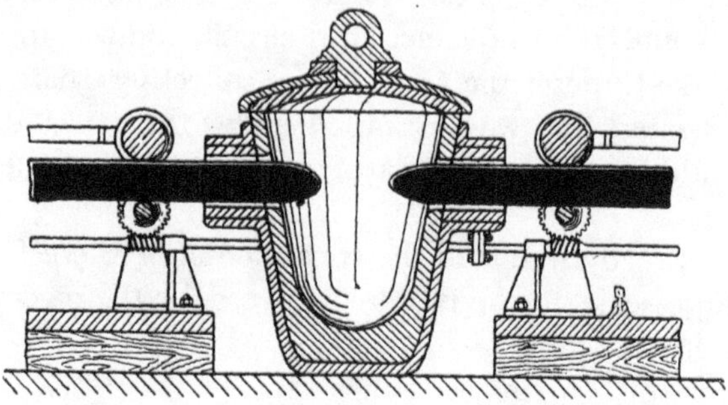

FIG. 177.

pp. 93–97, it can be established that Ch. W. Siemens had already in 1878 constructed furnaces in which—

1. The substance to be melted functioned as one pole of an arc, or

2. Was heated independently of the electrodes in a space heated by an arc; he had

3. Proposed and used electrodes fitted for cooling; and

4. Has understood how to influence the direction of the arc by the aid of electro-magnets in a manner most favourable to the utilization of the heat; finally, he has

5. Already effected the regulation of the electrodes by means of electro-magnets.

Indirect Arc Furnace of Rogerson, Statter, and Stevenson.—A furnace by Rogerson, Statter, and Stevenson[1] of the year 1886 also points to some originality. Instead of the Siemens closed solenoid, an open horseshoe electro-magnet is used here to direct the arc on the substance to be melted (Figs. 178 and 179).

Since then all that has been constructed with regard to furnaces belonging to this group may without difficulty be recognized as extensions and more or less further alterations in form of the furnace types already cited, with perhaps the exception of the constructions by *Girard & Street*, *Ducretet & Lejeune*,

[1] English Patent, No. 10,600 (1886); German Patent, No. 42,022.

Bertolus, Contardo, and Schuen, who have really created novelties in the points still to be considered below.

As we have already seen from the illustrations of two of his furnaces, Clerc had made the apertures for the carbon electrodes

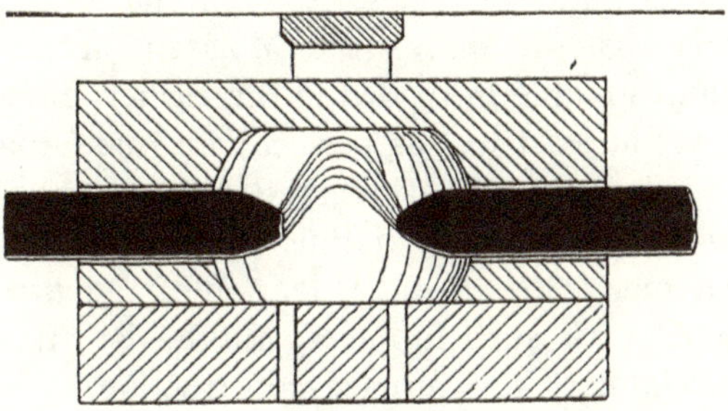

FIG. 178.

too narrow towards the interior of the furnace, which was probably due to the fact that he also wished to work the furnaces with explosive gas. It is, of course, well known that in an atmosphere of lime or calcium vapour the arc can be very largely extended. With normal arc voltages (65 volts), when the arc is about 3 to 5 mm. long in air, the ends of the electrodes can in

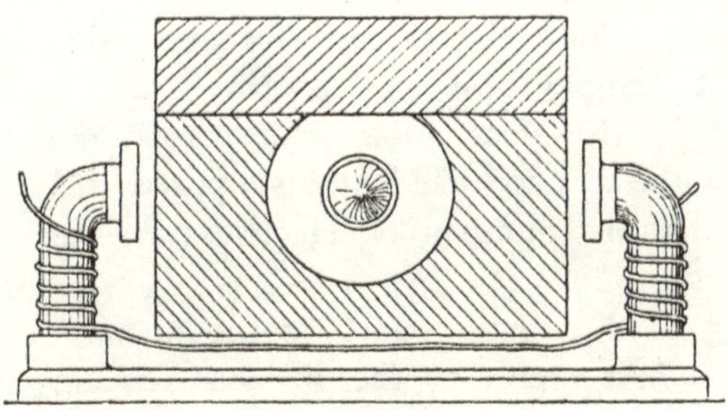

FIG. 179.

lime vapour be drawn apart to a distance of 50 to 60 mm. without disrupting the arc. As, however, the lime becomes fused and the magnesia decomposed, the narrow orifices very soon become so widened in the arc that the electrodes can be easily pushed into the furnace, as happened right from the start in the furnaces of Siemens, Rogerson, Statter, and Stevenson, and was kept in all the later imitations.

The Moissan Furnace.—The least deviations from the prototypes just mentioned are exhibited by a furnace of which Moissan styles himself the inventor. Other writers on electrochemical smelting processes have also accustomed themselves, through lack of a sufficient review of the literature of the subject, to denote this construction as the Moissan furnace. Nothing seems more unjustifiable in regard to the earlier furnaces of this group, which are at least just as perfect; for this furnace contains nothing in any way essential which had not already been applied in the earlier furnaces and had thus become known. Moissan deserves great merit for the investigation of the products which can be obtained with such furnaces, but he belittles it himself when he lays claim to the furnace construction as his own. It strikes one as peculiar when, in the official report of the Jury for the class on Electro-chemistry at the Paris Exposition of 1900,[1] it is stated that "The arrangement devised by Moissan has rendered it possible to exceed the temperatures hitherto attained in the industry by the aid of the various industrial furnaces (!). Those temperatures reached about 1800° C. until the oxyhydrogen blowpipe flame yielded 2000° C. In the Moissan furnace the temperature rose above 3000° C. (!)." Now, the Moissan furnace was not the first to reach these temperatures; in every older electric arc furnace using carbon electrodes such a temperature was possible, and had also been reached (and in the year 1902, when Moisson gave the first description of the furnace used and supposed to be invented by him, a considerable number of electric furnaces was already known).

According to the same report on the furnace, in agreement with Moissan's first report to the French Academy of Sciences, the following was stated:—"The arc springs across between two carbons pushed horizontally into a cavity inside a block of lime. A dome, also of burnt lime, closes the furnace. The substance to be worked is placed in a recess below the electric arc. The heating effect of the current, which is imparted to the lime in the immediate vicinity of the arc, is thus separated from its electrolytic action. This constitutes the characteristic feature of the

[1] "Rapports du Jury International, Class 24, Electro-Chemistry," p. 65 (Paris, 1901).

Moissan furnace (!)." *This constitutes the characteristic feature of all furnaces of this group, even the oldest!*

Furnaces of Moissan and Chaplet.—In some of the constructions of these furnaces Moissan[1] and Chaplet[2] used lateral feed-pipes for the material to be reduced, linings of carbon and

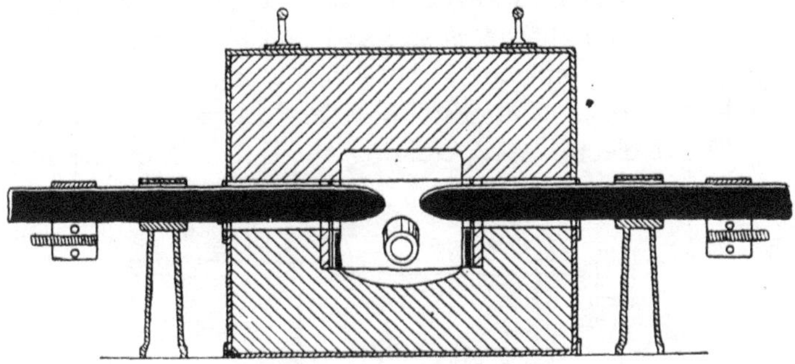

FIG. 180.

magnesia plates and different vessels, mostly pipes, for collecting the fused material. The hearths of these furnaces (Figs. 180 and 181) always consist of limestone blocks in which rectangular cavities are hollowed out. The walls of these cavities are lined with plates of magnesia about 10 mm. thick, and on these are carbon plates of the same thickness; carbon crucibles can also be inserted. Two massive carbons are introduced into the furnace through two of the opposite side walls. The carbon lining of the furnace space is, of course, cut out at these points so that the electrodes are not touched, and that at this part an arc cannot bridge across. At right angles to the electrodes, but somewhat deeper, a carbon pipe is brought through a third wall, and is arranged horizontally or inclined at an angle up to

FIG. 181.

[1] *Comptes Rendus*, **115**, 988 (1892); **117**, 679 (1893).

[2] German Patent, No. 74,537 of August 17, 1893, No. 77,896 of January 19, 1894; [English Patent, No. 15,577 of August 16, 1893].

30 degrees. The substance to be heated is in this pipe, or is conveyed through it into the furnace chamber. The end of the pipe should be about 10 mms. below the arc. The cover for the furnace consists necessarily again of a carbon plate, then a magnesia plate, and, finally, a limestone block.

In Chaplet's[1] first furnace the tube D (Figs. 182 and 183), which receives the charge, is so fixed that it can be charged and emptied from the outside. The furnace itself consists of the hollowed-out blocks of limestone A and B, through which are passed the arc electrodes C, crossing the direction of the heating

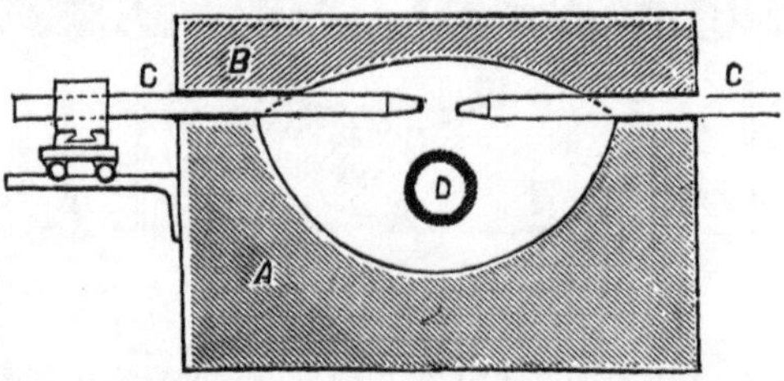

FIG. 182.

tube above the latter. Another method of arranging the furnace, with the muffle open on one side, is represented in Fig. 184.

Under the same name a second patent[2] has been granted

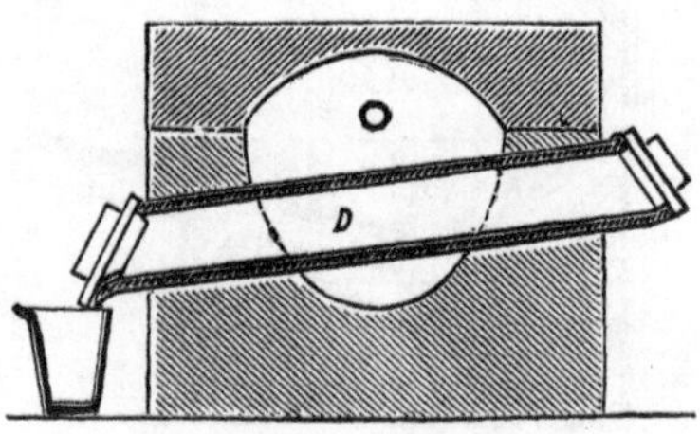

FIG. 183.

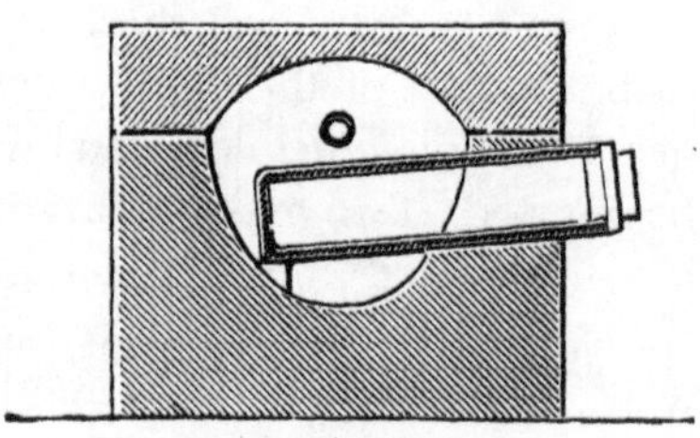

FIG. 184.

referring to the following construction. The hearth block *a* and the arched block *b* are made of similar material and in a similar manner to the furnace just described. In the block *a* several hollows or hearths, *e*, are made, and the arrangement is such

[1] German Patent, No. 74,532 of August 17, 1893; [English Patent, No. 15,577 of August 16, 1893].

[2] German Patent, No. 77,896 of January 19, 1894.

at after one hearth, or sump, has been filled the block *a* can e drawn aside sufficiently far beneath the block *b* for a new

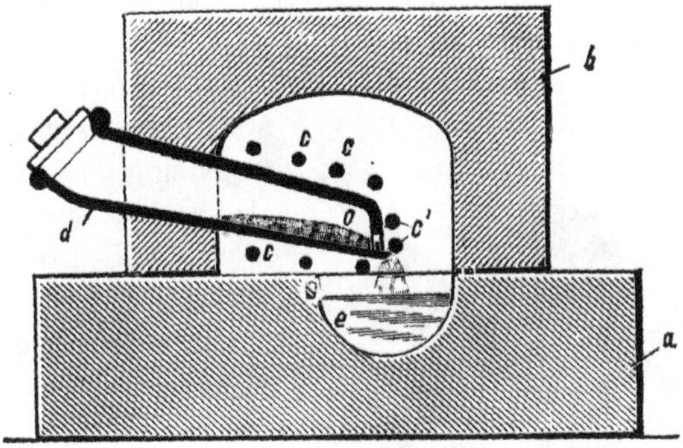

FIG. 185.

sump to come under the orifice *c* of the feed pipe, of which either one (*d*) or two (d_1 and d_2) are taken into the arch. The

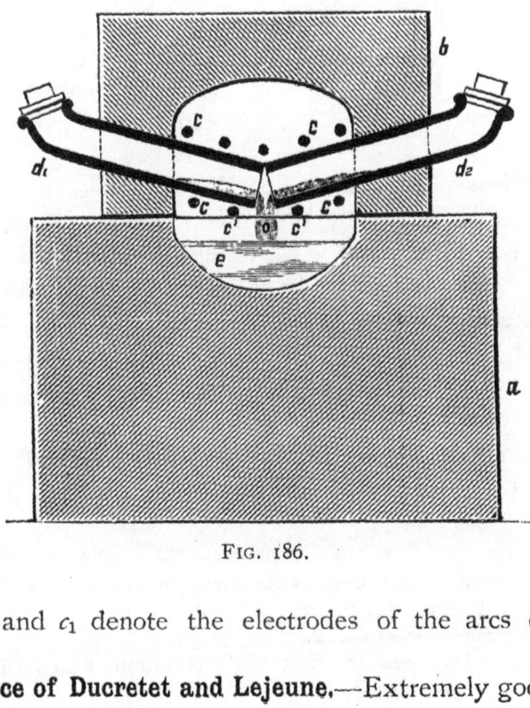

FIG. 186.

letters *c* and c_1 denote the electrodes of the arcs (Figs. 185 and 186).

Furnace of Ducretet and Lejeune.—Extremely good services for experimental purposes have been rendered by a furnace construction dating from this time (1892) by Ducretet and Lejeune.[1] In this furnace the atmospheric air can be excluded, and at the same time the process of smelting can be kept under observation. It contains openings for gas admission, two doors with mica plates, and a device for displacing, and if necessary also withdrawing, the smelting crucible from the heating chamber (Fig. 187).

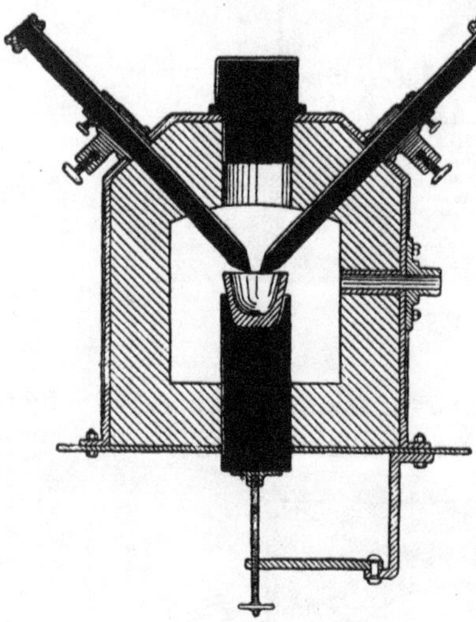

FIG. 187.

Zerener's Arc Soldering Apparatus.—As useful as Zerener's arc soldering apparatus,[2] in which the arc is diverted downwards by an electro magnet, proved itself for many metal working processes, so little was the hope fulfilled to utilize it like the blowpipe as a laboratory apparatus for preliminary experiments in the arc.

The Indirect Arc Furnace of W. Schuen.—In this respect a closed crucible furnace by W. Schuen[3] has given better results. The cover has fitted to it a closed electro-magnet, which serves at the same time as the handle. The electro-magnet also displaces the arc more powerfully and with greater certainty in the desired direction than an open horseshoe magnet. Two methods of construction of this very suitable experimental furnace are shown in Figs. 188 and 189.

The Zinc Furnace of the Trollhättans Elektriska Kraftaktiebolag.—For an electrical heating process void of any novelty

[1] *Electricien*, April 8, 1893 (Montpellier).

[2] "Verhandlungen des Vereins zur Beförderung des Gewerbefleisses," 1893; German Patent, No. 68,938; English Patent, No. 20,170 of 1894; "Jahrbuch der Elektrochemie," **2**, p. 113 (1895).

[3] Borchers, "Das neue Institut für Metallhüttenwesen und Elektrometallurgie," p. 47 (W. Knapp, Halle a. S., 1903).

patents[1] (de Laval) were also granted in Germany to the Troll-hättans Elektriska Kraftaktiebolag. It consisted in the appli-

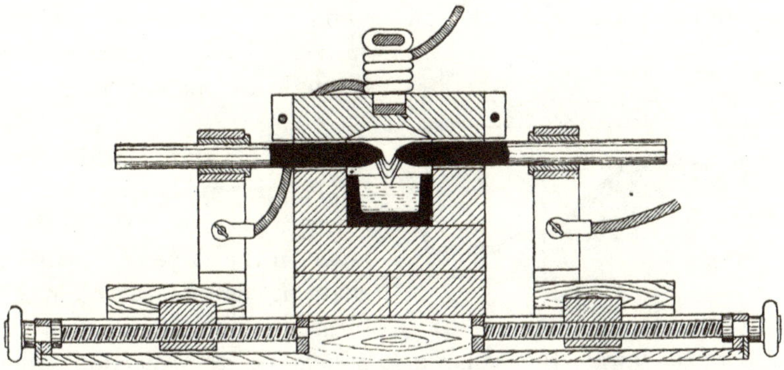

FIG. 188. Scale $\frac{1}{10}$.

cation of indirect arc heating for the distillation of zinc from the surface layer of the slope of a heap of raw material which is

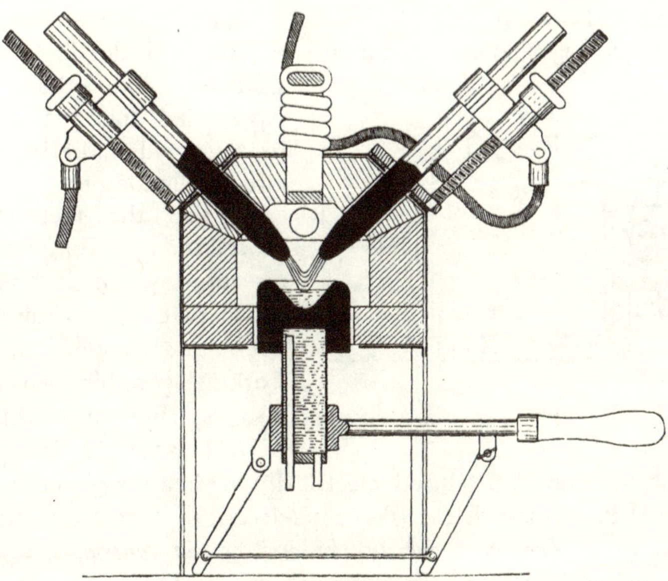

FIG. 189. Scale $\frac{1}{10}$.

[1] German Patents, No. 148,129 of December 19, 1901 ; No. 148,439 of April 30, 1901 ; and Supplementary Patents, Nos. 157,603 and 162,535.

pressed into the furnace chamber. The arrangement of the electrodes and of the charge is clear from Figs. 190 and 191.

By means of two supplementary patents the idea was then also protected to allow the gases and vapours evolved on the surface of the slope to escape through an opening free from the charge (!).

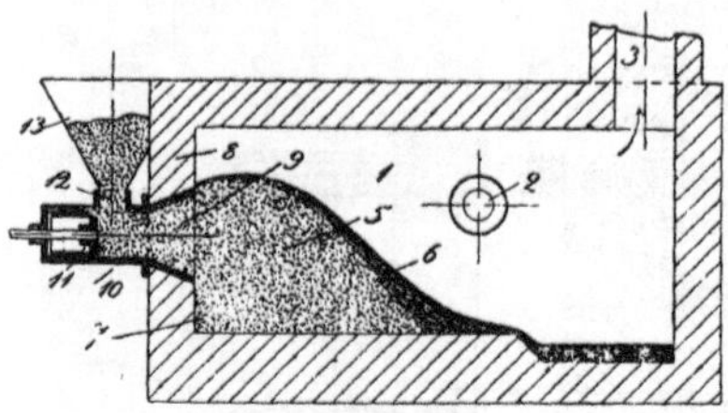

FIG. 190.

To melt and volatilize materials of any description by electrically heating the surface layer on the slope of a heap of charging material placed in the furnace chamber, as well as to conduct away the volatile products of the reaction from electrical furnaces through an opening free from the charge, constitute frequently applied principles in the construction and working of electric furnaces (Cowles, Readman and Parker, Price, King and Wyatt, Rathenau, Deutsche Gold- und Silberscheide-anstatt, and others).

In working zinc ores in furnaces of this kind, the Trollhättan Company, like others who have tried this method of heating, will make, or have already made, the experience that the method of procedure according to the directions given in the de Laval patents unfortunately does not render smooth working possible, which is also confirmed in the reports of J. Hess: "In Hasslund (Norway) zinc is produced electrically with an expenditure of 2400 H.P. by the Elektriske Smaeltverk of Gustav de Laval. *Certainly, at present, working the ores is not contemplated*, and only spelter is worked up electro-thermically to a product containing over 99·9 % of zinc. It is well known that with the ordinary process it is only possible to obtain refined zinc containing at most 98·8 % of zinc, as the remaining 1 % of lead

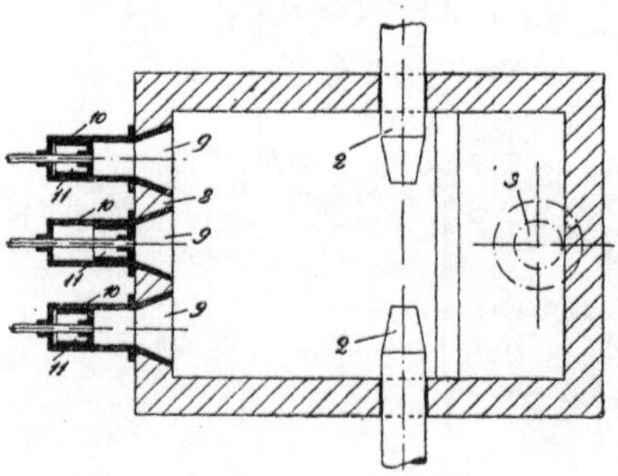

FIG. 191.

cannot be reduced. A plant is to be installed for the Trollhättan's Elektriska Aktiebolag for working zinc sulphide from Saxberg, near Ludwika."[1]

Patin's Indirect Arc Furnace.—A furnace designed by Patin[2] with hermetically sealed cover for working under pressure, or *in vacuo*, has only been used for a few experiments at Puteaux, near Paris; it was soon replaced by an open pit-furnace such as has been described under the furnaces with movable hearths.

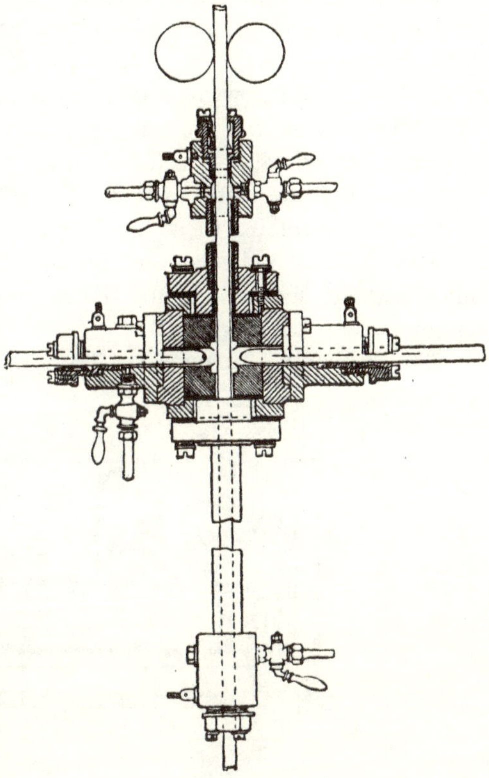

FIG. 192.

Indirect Arc Furnace of Girard and Street.—The Patin furnace was, moreover, in its essential parts, only a copy of the furnace by Girard and Street[3] for the graphitization of carbon bodies. Of the numerous possible constructions described and illustrated in the patent specifications of the inventors, the following has survived in practice, as I am able to confirm from my visits in Paris to the Le Carbone Works:—

The carbon rod is first compressed with the graphitizing

[1] "Jahrbuch der Elektrochemie," **10**, p. 710 (1903).

[2] German Patent, No. 94,641 of September 12, 1896; [English Patent, No. 19,290 of September 1, 1896].

[3] English Patent, No. 13,340 of 1893, No. 15,632 of 1894; German Patent, No. 78,237 of 1894, No. 81,479 of 1895, No. 85,335 of 1895; United States Patent, No. 571,655 of 1896; Borchers, "Elektrochemie auf der Pariser Weltausstellung," p. 91 (W. Knapp, Halle a. S., 1900).

ingredients and baked in the customary manner. It is then slowly passed through a narrow heating chamber contained in a magnesia block, while the chamber is heated by an arc from lateral electrodes. For uniformly spreading the arc rays over the glowing body as it passes through, solenoids were at first proposed by Girard and Street. The working soon proved, however, that, when the heating chamber had reached a sufficiently high temperature, a uniform distribution of the current took place together with a uniform heating of the carbon body. Whereas, at first, the arc only springs across to the cold carbon rod at isolated spots, utilizing the rod mainly as an intermediate electrode, with increasing conductivity of the chamber walls and its contents these local actions cease.

The carbon bodies enter through a long conducting pit in uninterrupted succession, and leave the furnace through a still longer cooling pit (Fig. 192).

The Roberts Furnace. — The belt conveyor hearth of

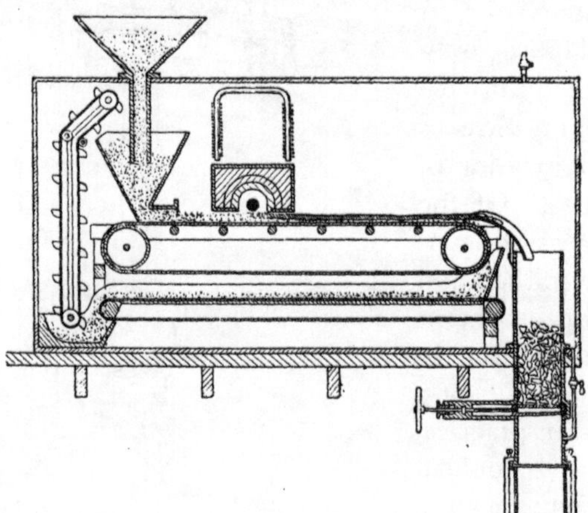

FIG. 193.

Roberts,[1] which has already been mentioned among the resistance furnaces, can, of course, be adapted for arc heating. In this case right over the belt which conveys the charge is built an electric reverberatory furnace, the arc of which is situated

[1] United States Patent, No. 588,012; [English Patent, No. 17,609 of December 29, 1896].

beneath an arch of limestone, powdered lime, masonry, and iron fastenings, and is directed on the material to be fused by means of an electro-magnet. The charge is conducted and delivered through hoppers. As it is being carried forward by the belt conveyor, its height is so regulated that in its passage through the zone of the arc only a portion of the mass is melted. In this way the underlying part will protect the conveyor belt from the high temperatures. Also by means of a scraper the solidified product is passed into a collecting and cooling chamber, from which it can be tapped from time to time. A belt conveyor carries the unmolten mass into a collecting pit, from which it is again raised by a bucket elevator. The whole is enclosed, so that by the aid of valves any desired gaseous atmosphere can be kept in the heating chamber (Figs. 193 and 194).

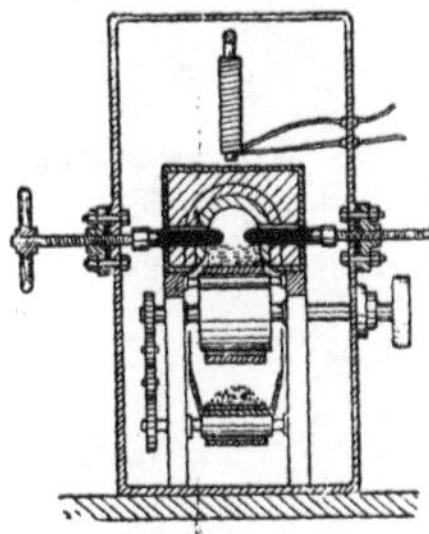

FIG. 194.

The Bertolus Multiphase Current Arc Furnaces.—For the direct application of multiphase currents Bertolus constructed a new furnace which has as many electrodes as there are phase currents. The furnace chamber is constructed of masonry, and is lined internally with carbon. The charge is fed in laterally through hoppers. There are, besides, working openings left free in the charge walls. A dam regulates the outflow of the fused material into a breast-pan, or collecting vessel. The furnace is covered by a ring, through which the electrodes are passed, and by a cooled gas-trap. The electrodes are fastened to guide-rods, and are here connected to the supply cables. The guide-rods are connected to moving mechanisms, so that each electrode can be separately adjusted or all of them simultaneously (Fig. 195).

According to estimates by Bertolus, with the same densities the output of the furnace should exceed that of direct current furnaces as many times as there are phase currents; or, if only the same output be required, the current density worked with should be considerably less. Thus the parts mostly exposed to wear, the electrodes, their connections and moving mechanisms can be built much lighter and be more easily controlled.

Bresson's Furnace.—In the second[1] furnace by Bresson[2] furnished with a smelting hearth of the tilting type, the electrodes are introduced laterally. But the tipping-gear and the mechanism

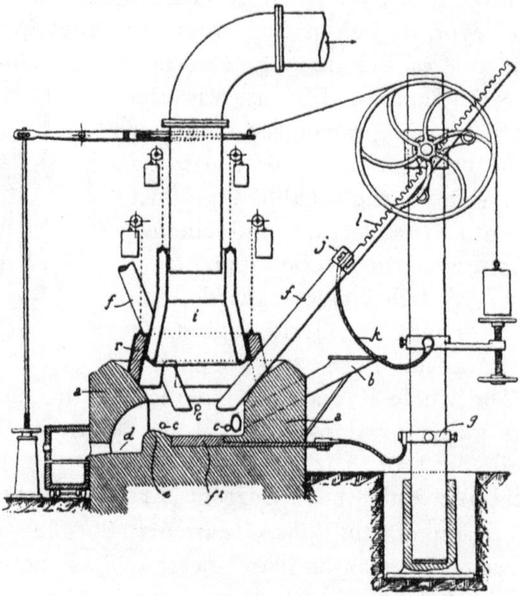

FIG. 195.

for moving the electrodes combine together to such a complicated apparatus that the furnace does not commend itself either for practical working or for experimental purposes.

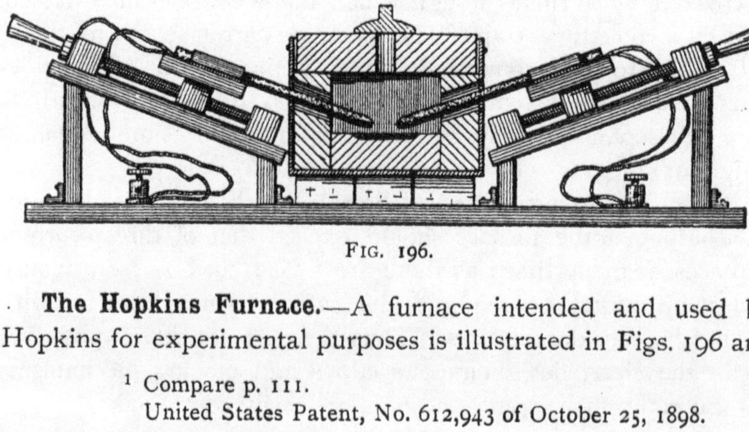

FIG. 196.

The Hopkins Furnace.—A furnace intended and used by Hopkins for experimental purposes is illustrated in Figs. 196 and

[1] Compare p. 111.

United States Patent, No. 612,943 of October 25, 1898.

197, and a diagram of connections is given in Fig. 198. The working of the furnace follows at once from these illustrations. It is also clear that the number of the electrodes was chosen according to the available voltage of a supply circuit in the

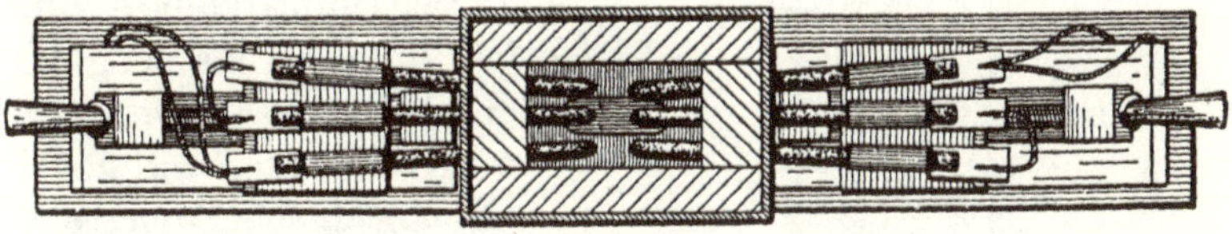

FIG. 197.

Hopkins laboratory, and that in other laboratories other current conditions must be met by alteration of the number of the electrodes.

Contardo's Furnace.—In the electric furnace by Contardo,[1] for the purpose of effecting a preliminary heating of the charge, the latter is passed over a hood arranged over the furnace chamber

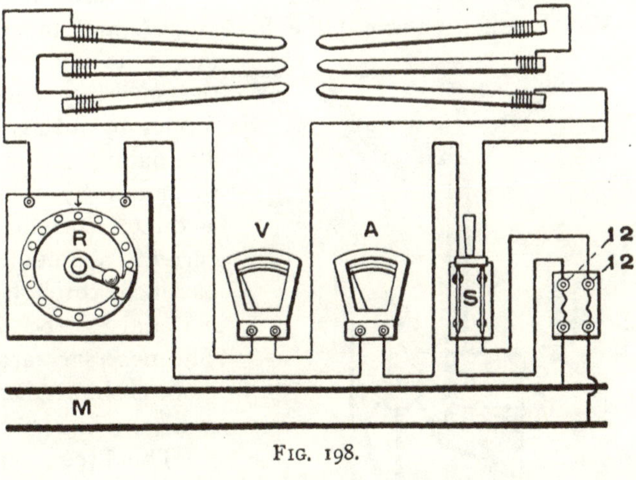

FIG. 198.

as the waste gases escape in the opposite direction. In order to still further utilize the waste gases for this purpose, they are led into a combustion chamber in which the flame circulates round obliquely arranged pipes, through which the charge is carried

[1] English Patent, No. 4575 of 1901; United States Patent, No. 750,753 of January 26, 1904.

in before it reaches the cover over the mouth of the furnace. Contardo also repeats the proposal, already several times expressed, to use the waste gases of electric furnaces—mostly very rich in carbonic oxide—for driving gas engines after the gases have been cleaned (Figs. 199 and 200).

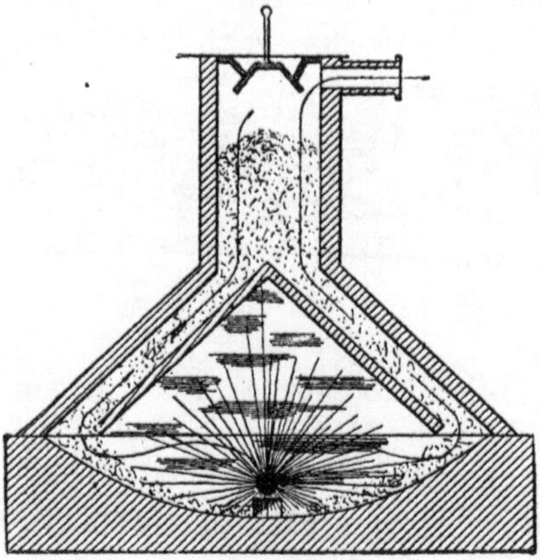

FIG. 199.

Von Seemen's Furnace.—Von Seemen's process[1] for handling ores, metals, etc., in electric furnaces differs from those furnaces which have hitherto become known in which the heat required for smelting is supplied from outside and the reduction is effected by a direct current flowing through the fused material. In the Von Seemen's furnace the heat supply and the reduction take place inside the smelting space, the former by means of an alternating current, and the latter by means of a direct current, both currents simultaneously passing through the material to be fused. The necessary arrangement is shown diagrammatically in Fig. 201.

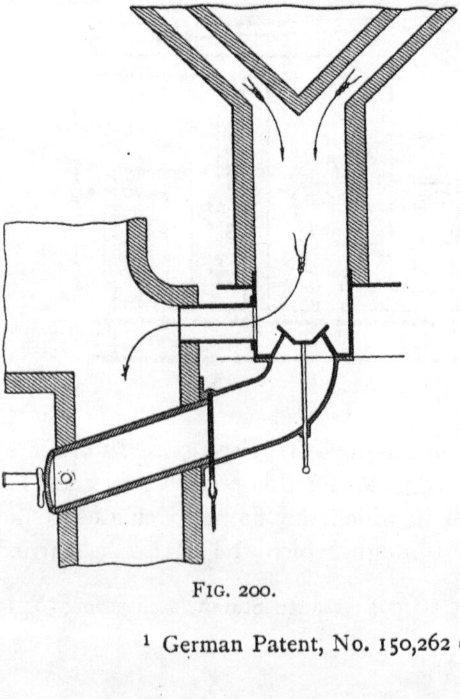

FIG. 200.

The three electrodes *a* are inclined towards a point *b* of the smelting chamber at a definite angle which has to be determined by experiment, and are

[1] German Patent, No. 150,262 of June 12, 1903.

connected to the terminals *k* of a multiphase alternator. Arranged between them in the centre is the electrode *d*, also directed towards the point *b* and joined to the one pole of the direct-current dynamo. A choking coil *e* interposed in the circuit prevents the intermingling of the direct and multiphase currents. The point *b* stands in communication with the other pole of the direct-current dynamo and with the neutral point *c*. The circuit running from *b* to *c* is provided with a switch *f* to allow the multiphase current to go from electrode to electrode, or from the electrodes to the neutral point; it is also provided with a condenser *g*, to prevent, as above, the intermingling of the two current systems.

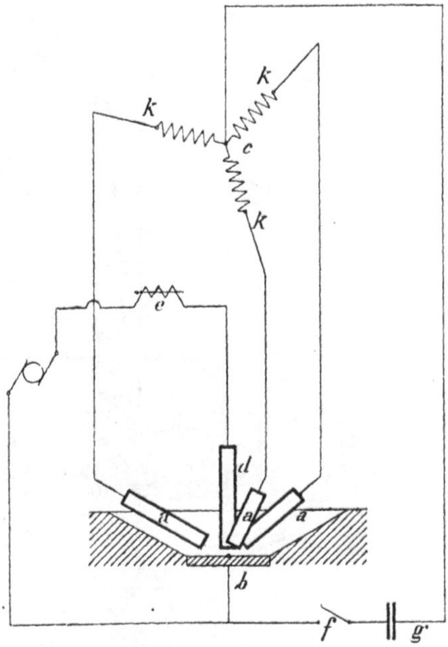

FIG. 201.

This arrangement of the electrodes and circuits enables the material to be converted to the liquid state by multiphase or single-phase currents, and to be reduced by direct current, as both kinds of current can be simultaneously taken through the charge.

The Arc Furnaces of Stassano.—Since the end of the last century Stassano[1] has been attempting to construct suitable furnaces for smelting pure iron ores so as to produce malleable varieties of iron. His first electric furnace was made as a pit furnace, and was, therefore, described in the reports of the technical press as an electric blast-furnace. This furnace has certainly been constructed and put to work in connection with a water-power installation in the Camonica Valley of the Bergamo Alps in North Italy, although any one with any experience of

[1] English Patent, No. 11,604 of 1898; cp. also "Jahrbuch der Elektrochemie," **6**, p. 320 (1899).

electric furnaces must have been convinced from the start of the futility of this attempt (Figs. 202 and 203).

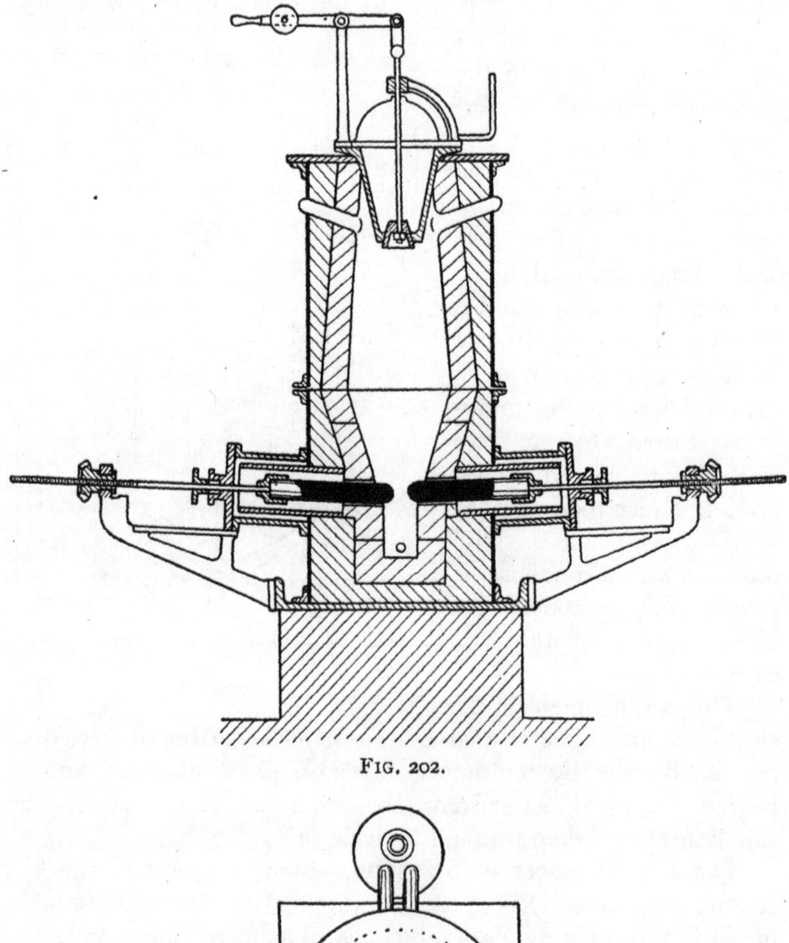

FIG. 202.

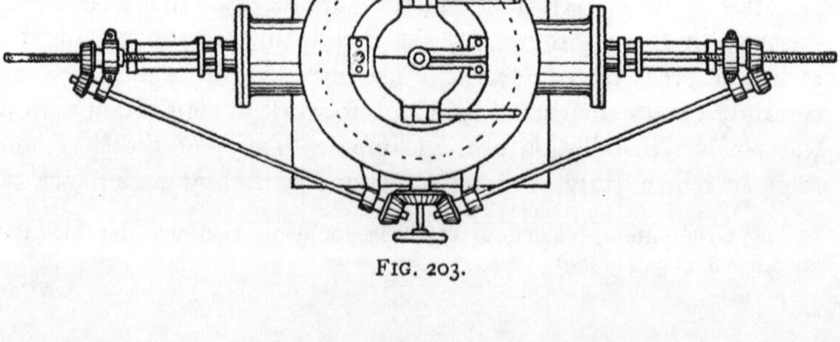

FIG. 203.

Apart from an alteration of this furnace, which we can pass over, he then constructed it in the form of a reverberatory hearth furnace, in which several pairs of electrodes, and consequently several arcs, were distributed above the molten masses (Figs. 204 and 205).

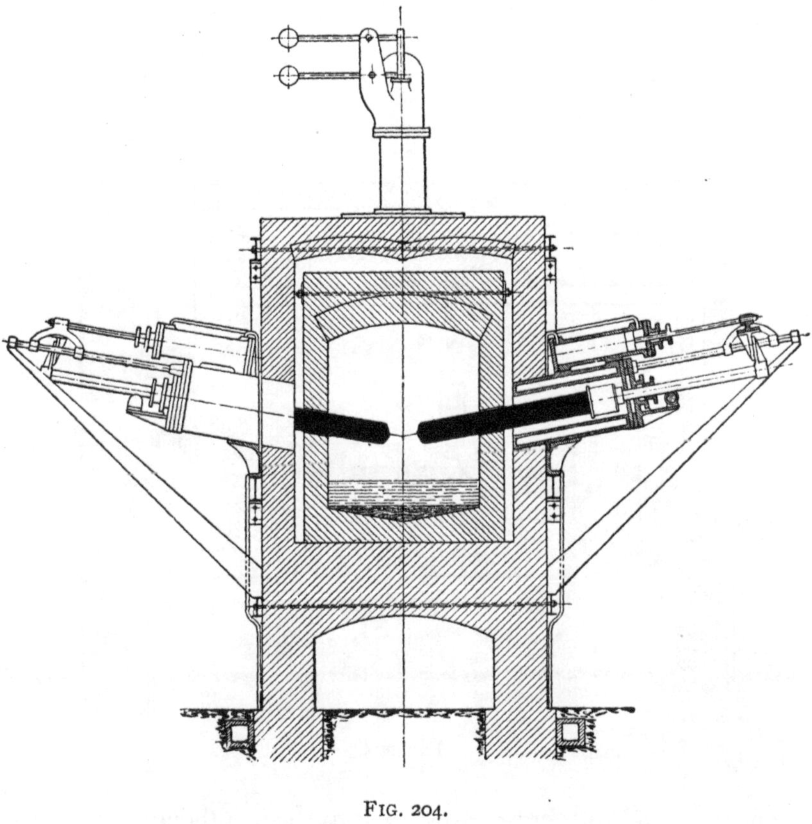

FIG. 204.

Also this furnace Stassano finally replaced by a construction having a circular hearth situated in a low cylindrical pit, in which two or more electrodes project above the level of the fusion. For more thoroughly blending the materials of the charge, the furnace was erected on rails slightly inclined to the vertical, and was slowly rotated (Figs. 206, 207, 208).

With this furnace Stassano believes to be able to produce

6 (metric) tons of malleable iron with 1000 electrical H.P. direct from very pure iron ores.

The Mercury Vapour Lamp.—In this section of indirect arc heating I should not omit to mention those devices in which

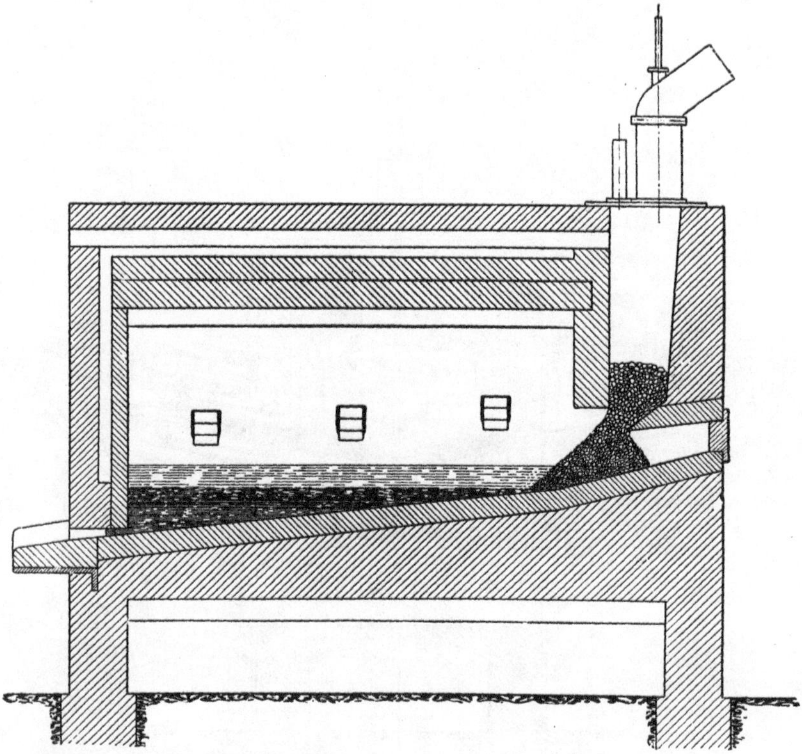

FIG. 205.

different kinds of actinic rays are produced, although it will be agreed with me that the majority of them cannot be regarded as electric furnaces in the sense of the word given in this book. I will therefore restrict myself to one example, which, on the one hand, offers the possibility of a closer study, and, on the other hand, is destined to become of importance in metallurgical investigations on account of the large quantity of ultra-violet rays emanating from this arc, namely, the mercury vapour lamp.

The Heräus Mercury Vapour Lamp.—From the Platinum

Refinery of W. C. Heräus I have received the following report of two of their mercury vapour lamps :—

In the one lamp the arc is produced in the quartz glass tube *a* (Fig. 209). This tube, in the normal lamp, has a diameter of about 9 mm., and is curved in the form of an S. The mercury electrodes are contained in the two vertical tubes which, for cooling purposes, are fastened to the metal cheeks *c*. The tube at *b* serves to automatically ignite the lamp. It is filled with mercury, and is covered with a cylinder of sheet asbestos on which a platinum wire is spirally wound. This cylinder acts as the electric heating apparatus. It generates mercury vapour in the tube at *b*, the vapour pressure forces the mercury into the combustion tube *a* up to the +pole, the heating current is automatically interrupted, and the mercury vapour at *b* condenses again and sucks the mercury back into its original position, whereby the arc is formed.

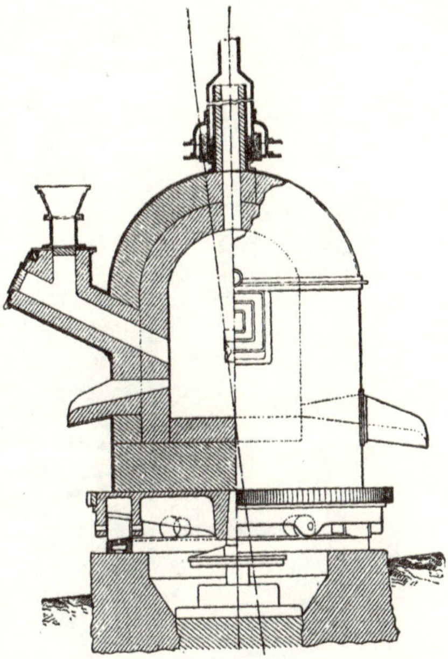

FIG. 206.

A small mercury lamp of quartz glass suitable for demonstrational purposes is shown in Fig. 210.

A pear-shaped glass vessel about 5 cm. wide and 8 to 10 cm. long is fitted at the top and bottom with tubular attachments for the two mercury poles. The negative pole is at the top, and the positive at the bottom. On the outside the negative pole is provided with a cooling arrangement. The current is conducted to and from the apparatus by means of split nickel-steel pins with a layer of mercury on the outside. This layer of mercury is secured by a cement. Before starting the lamp the top electrode tube is completely filled with mercury, and the lamp is then fixed vertically in the stand. The lamp can be connected to an electrical heating circuit of 65 to 220 volts through a resistance the value of which depends upon the particular voltage. With 65 volts a resistance of about 15 ohms should be used, with 110 volts

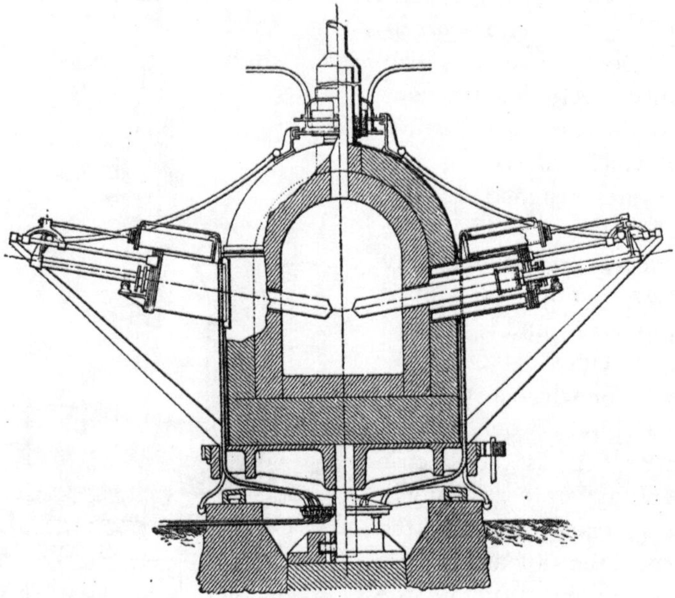

FIG. 207.

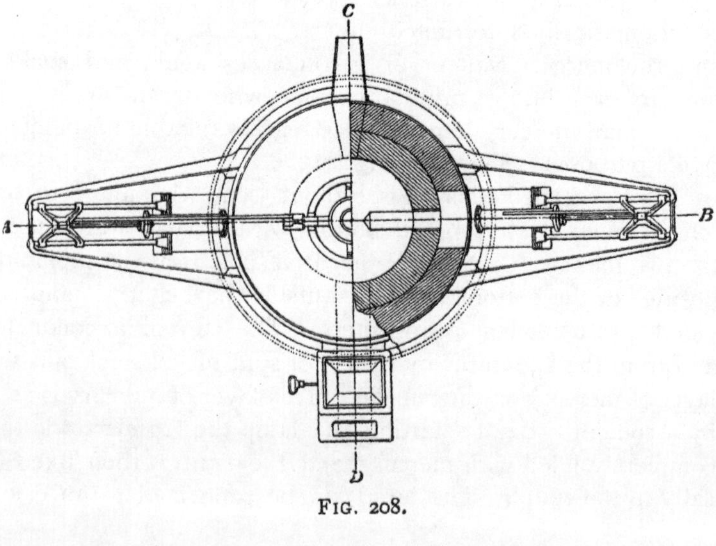

FIG. 208.

30 ohms, with 220 volts 95 ohms. The current is 1·5 to 2·5 amperes.

A small induction coil is used to ignite the lamp. One suffices having a spark-gap of 7 to 10 mm., which can be worked with a small accumulator. The one terminal of the secondary coil is connected to the negative terminal of the lamp, the other terminal to the metal cooling device of the negative pole. The lamp

FIG. 209.

ignites as soon as the current connections have been properly made and the primary circuit of the induction coil has been momentarily closed by a key. It is advisable to let the spark of the induction coil also spring across a small parallel spark-gap, as in this manner the ignition is surer, and the lamp, on the other hand, is secured against rupture of its walls by too strong a spark. In case the ignition does not take place at once, slightly warming the upper part of the lamp will accomplish this, especially the attachments of the negative-pole tube. With too large a resistance in series with the lamp, it easily happens that although the lamp ignites it goes out again soon afterwards. In this case the ignition must be repeated several times, or the resistance slightly decreased.

After ignition the pear-shaped part of the lamp appears quite full of pale, diffused light, which is characterestic of a low vapour density. In consequence, the smell of ozone becomes specially strongly marked after ignition. The mercury which is vaporized from the electrodes condenses on the walls of the tubes. Local heating of the walls with a Bunsen burner will then force the arc away from the wall to the opposite side through the mercury vapour which forms. With the increased heating of the walls

FIG. 21C.

caused by the vaporization of the mercury which has condensed there, the vapour density inside the lamp rises, and the column of light contracts axially to a constantly diminishing streak. With an arc burning like this the deviation produced by a magnet can be demonstrated in a very characteristic manner. The arc is bent into a curve tangential to the lines of force of a horseshoe magnet, and can, finally, be extinguished. The corresponding changes which occur in the current and pressure can be easily shown with lecture instruments.

If the diffused arc light is to be maintained for longer periods more resistance must be switched into circuit after ignition.

The ultra-violet rays produce a very painful inflammation of the eyes. The eyes should, therefore, be protected by spectacles as soon as the lighted lamp is approached. With long exposures the skin also becomes inflamed. Observations are, therefore, best conducted with a glass screen which absorbs the ultra-violet rays.

CHAPTER VII

FURNACES ARRANGED FOR DIFFERENT MODES OF HEATING

Borchers' Experimental Furnaces.—It is very evident that a whole series of furnaces can be used without alteration, both for resistance heating and for arc heating, by allowing both the electrodes to dip in the charge in the former case, and only one or none in the latter. For furnaces intended for practical working this range of application will in most cases be found sufficient; the furnace is constructed for a certain process mainly with the one object of keeping down the capital and working costs. For experimental purposes, however, it matters less to utilize the current to the best advantage than to be able to easily and quickly change the conditions of testing. With this object in view I designed a series of furnaces for the Electro-Metallurgical Laboratory, under my direction, at the College of Aachen, so that by changing individual parts they could be easily altered for any mode of heating. The first furnace was intended for experiments on a small scale. At the time when it was made I only had 12 to 20 electrical horse-power at my disposal.

The construction and the method of working the furnace for direct resistance heating have already been considered on pp. 14 to 15. It is, therefore, only necessary for me to repeat the illustration (Fig. 211) for the purpose of comparison.

For indirect resistance heating the electrodes are introduced as in the direct resistance heating method after the stone walling has been put in. A thin carbon rod is then placed between the two electrodes, and round it is packed the charge to be subjected to the heating (Fig. 212).

When with heating processes of this kind volatile products,

which can be condensed again, are to be expected from the reaction, then, of course, care must be taken to seal the smelting-space as efficiently as possible. For instance, in his experiments on the reduction of zinc ores rich in silicates, *Dorsemagen* amplified

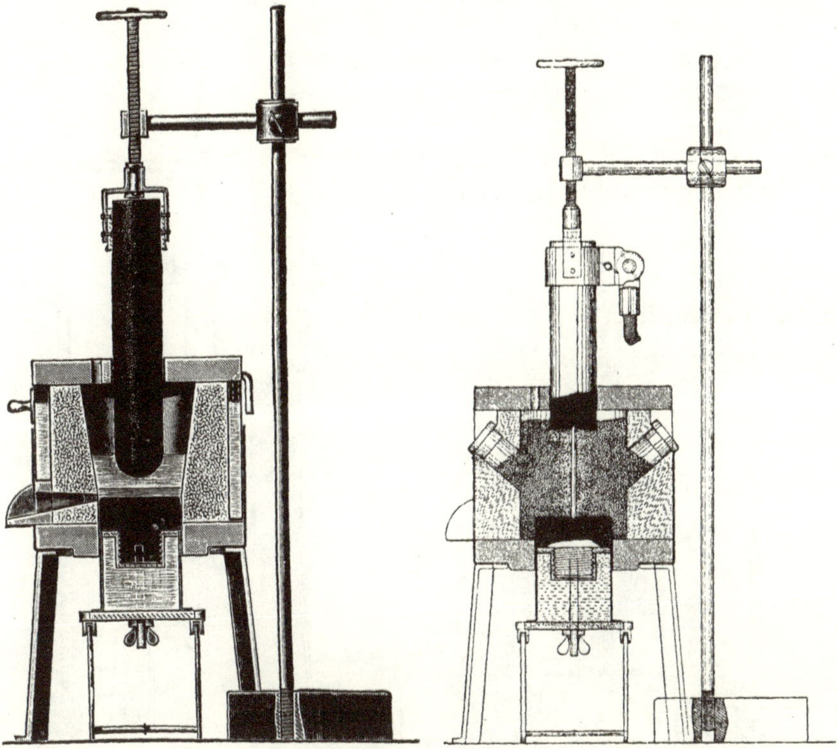

FIGS. 211, 212. Scale $\frac{1}{10}$.

the furnace parts just described by the following arrangement: A crucible from which the bottom had been removed to admit the lower electrode was placed round the electrodes so as to receive the ore to be melted. To this crucible was again connected a smaller one of clay by means of a short iron pipe to act as receiver. The joints around the electrodes were made tight with a little asbestos, and the rest by filling in the spaces between the crucible and the furnace lining with fine sand; the upper part of the furnace has a sheet-iron crown to hold the sand packing. The arrangement is shown in Fig. 213.

For direct arc heating the arrangement of the electrodes remains the same. Instead of a block, a carbon crucible is more advantageously used as the lower electrode (Fig. 214). The upper electrode should be selected as thin as is permissible with reference to the available current. The furnace charge in this case is best introduced through a laterally placed carbon tube after the furnace has first been heated by the arc.

For indirect arc heating both electrodes should be introduced at the side. The lower electrode holder now only plays the part of a crucible holder. Of course, a flat smelting crucible without this holder can be put in the bottom of the furnace. In this case also the furnace charge is introduced through a lateral carbon tube (Fig. 215).

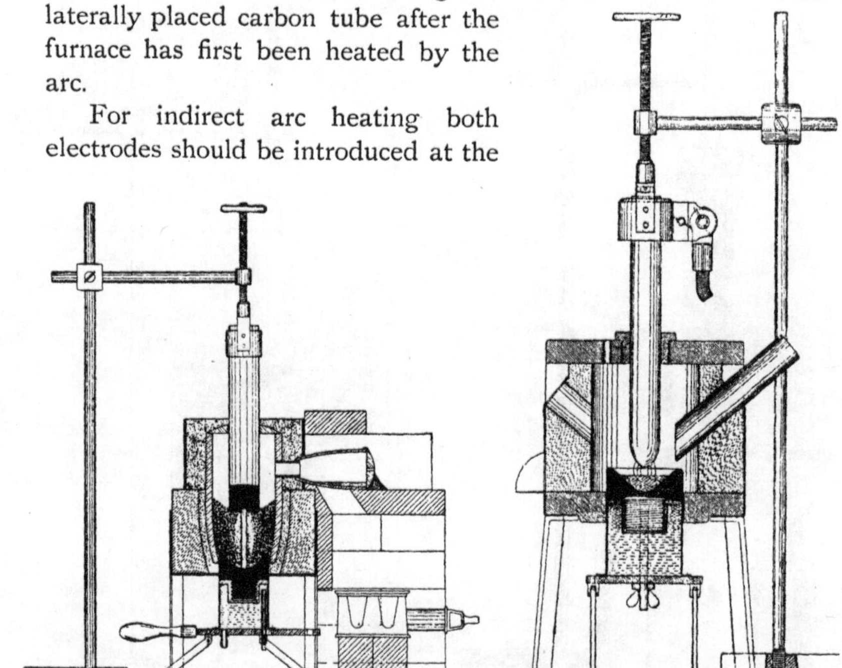

FIGS. 213, 214. Scale $\frac{1}{10}$.

When in 1898 I exhibited this furnace at the German Electro-Chemical Society, opinions were divided whether the electrode holders should be rigidly connected to the frame of the furnace, or not. A few were in favour of the former arrangement, which for good reasons (less convenient handling of the furnace itself, more difficult alteration of the heating system, and greater insulation difficulties) I did not choose for these small apparatus. When, in the year 1901, the question arose as to the arrangement of such experimental furnaces on a larger scale (60 to 100 H.P.) for the equipment of the new Institute of Mines and Electro-

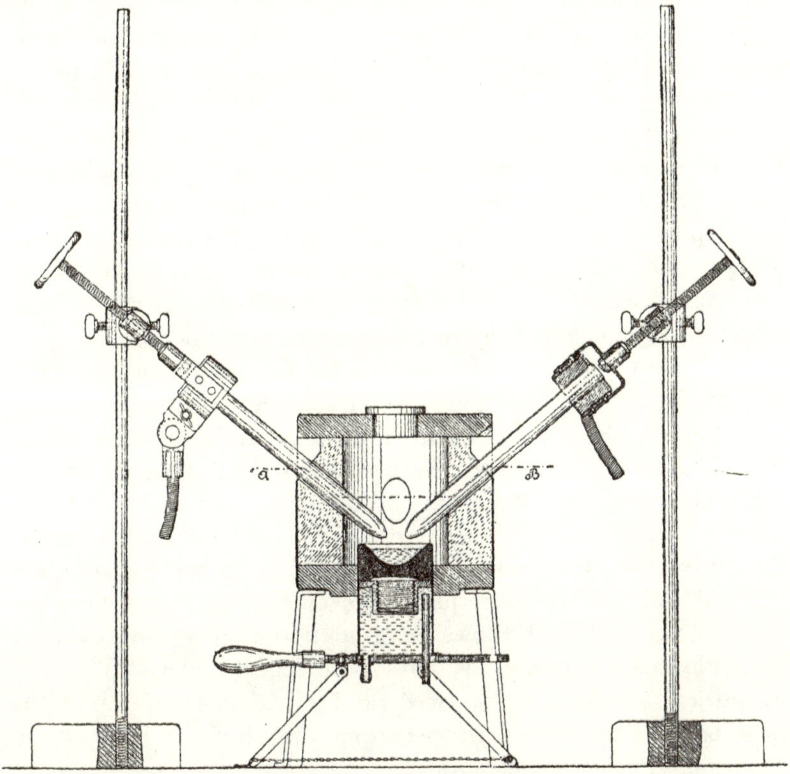

FIG. 215. Scale $\frac{1}{10}$.

Metallurgy, I considered placing the electrode holders on the frame of the furnace as exceedingly suitable in this case, on the assumption that with these dimensions they would be more easily handled when carried out in this way, and would render the furnace more accessible than with special frames. The result was the following furnace carried out partially on the lines proposed by the mechanician, Mr. *W. Schuen*, of Aachen :—

The various parts of the furnace are to a certain extent

suspended on two spindles, which are horizontally mounted in a framework built up of channel iron. By means of a handle the spindles can be turned in any manner, and they can be rigidly held in any position by clamping a spring clip of hoop iron against a fixing-ring. In the middle these spindles support a ring which is also made of channel iron. The ring is open in the centre, and is cut out at the top so as to receive the different furnace hearths. Near this ring-carrier two pairs of thin pillars are rotatably mounted on the spindles by means of bosses, and in turn carry the holders and guides for the electrodes. The electrode holders by means of which connection is made with the current supply are first pushed on the pillars; they consist of a clamping piece shaped like a shaft bearing, to which two lateral arms connected to the guide sleeves on the pillars are joined. The guide sleeves are oval-shaped. On the side arms of these holders are the contact sleeves for the current connections.

Although the electrodes should be firmly clamped in their holders for the sake of making good contact, the latter are really supported by the outer cross-arms of the pillars. As Figs. 216 to 219 clearly show, hollow screws are inserted in the terminals of the electrodes, and are easily adjusted by the screwed hand-wheels carried on the outer cross-arms. As far as the screw-thread extends the screw forms a complete hollow cylinder, the internal width of which is just large enough so that it does not touch the carbon. The lower continuation of this screw cylinder is made narrower, and is cut to a half-cylinder. The two cylindrical halves are, of course, rigidly held in the clamp of the electrode. The method of insulating the electrode holders and guides is sufficiently clear from the sketches.

On the one side of the furnace framework, on the right in Figs. 216 and 217, a second stop-ring is fastened to the supporting frame, by means of which the rotatably mounted electrode holders can also be firmly held in position. As here there was no room for the screw device used with the first stop-ring, this was carried out by means of pins, which fit in suitable holes of the stop-ring and electrode arms. Figs. 216 to 219 clearly illustrate the positions which the electrodes can be made to take up by the aid of this mechanism.

By simply changing the smelting hearths, always using the

same electrode frame, furnaces for all the modes of electric heating can be made. The hearth depicted in Fig. 216 is principally intended for effecting electrolysis of the fused salt;

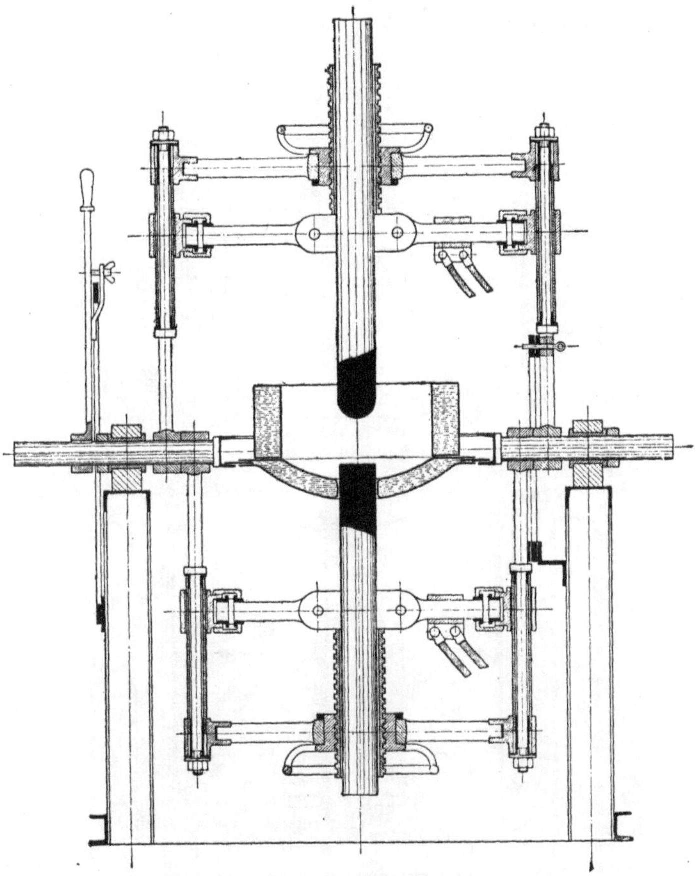

FIG. 216. Scale $\frac{1}{20}$.

it consists of two water-jackets, of which the lower one is in the shape of a flat bowl, while the upper one forms a short cylinder. After the bottom electrode has been introduced and is in the position shown in the drawing, a hearth is then formed exactly as in the small experimental furnace described for the extraction of aluminium. The same material is used, which should then be melted and electrolyzed.

For electrolysis in the molten bath the arrangement shown in Fig. 217 may also be used. The cooled floor of the hearth described above has been retained, while the cylindrical water-jacket is replaced by a cylinder of carbon. This construction is utilized when it is required to produce as high a current density

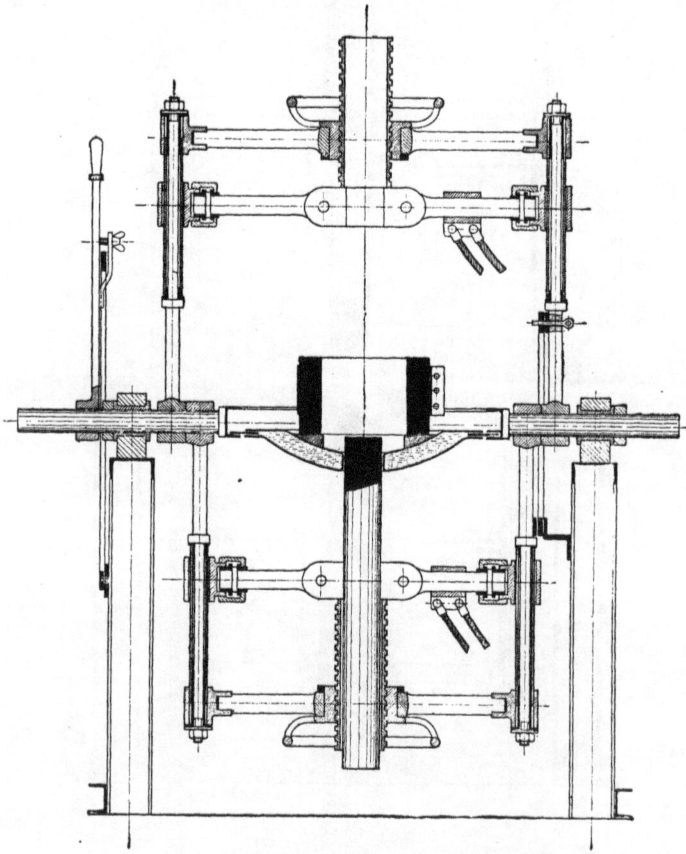

FIG. 217. Scale $\frac{1}{20}$.

as possible at the cathode. The carbon cylinder, which, as in the small apparatus (cp. pp. 44 to 46 and Figs. 48, 49, 50, and 52), is composed of segments like the staves of a cask, naturally forms the anode.

Fig. 218 shows an arrangement for direct arc heating. On the lower electrode stands a crucible of conducting coke, into which the upper electrode can be introduced vertically. The crucible is placed in a sheet metal case lined with a bad heat-conducting material to avoid heat losses as far as possible. The case is packed so that volatile substances can be distilled through the opening in the cover shown on the right in the illustration.

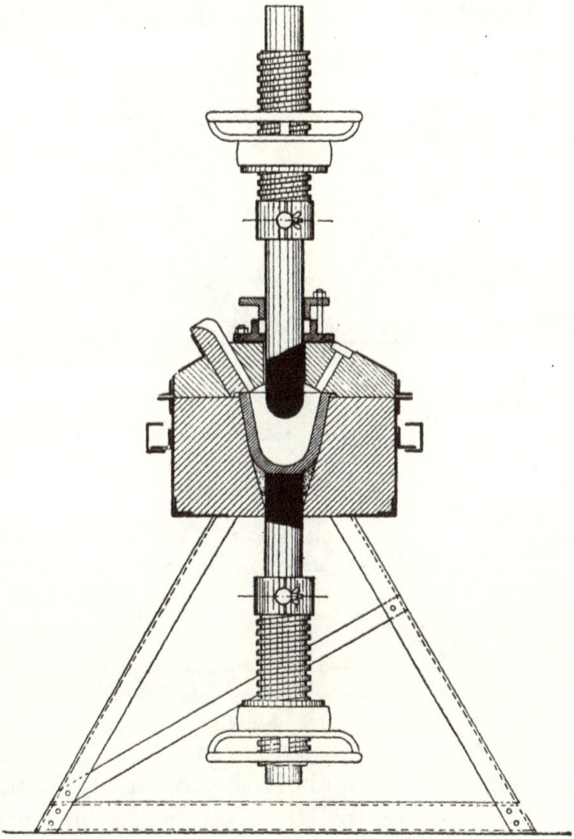

FIG. 218. Scale $\frac{1}{20}$.

The opening in the cover on the left is for the purpose of charging the furnace during the working, and for pouring out the fused mass when the experiment is finished. The latter operation can take place without having to dismantle the apparatus by revolving the bearing pivots with everything on it.

Fig. 219 shows two positions of the electrodes in a fourth

smelting hearth. The dotted position would be the one for indirect arc heating, whereby the arc forms between the two electrodes and heats the furnace charge by radiation. The charge is introduced through an opening in the centre of the cover. In the position of the electrodes drawn out in full the furnace would be applicable for the new heating process of Héroult. According to this process the electrodes introduced

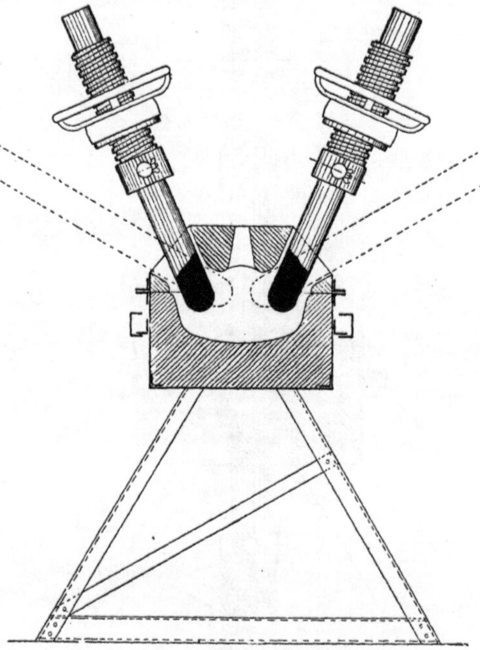

FIG. 219. Scale $\frac{1}{20}$.

into the smelting vessel from above should not dip in the actual smelting product (metals free from carbon, as iron, chromium, manganese, and so on), but in a layer of slag, or chemically active fluxes, to be kept above it, so that the electrodes do not come into direct contact with the metal. Only as thin a layer of slag or flux should exist between the electrode and the metal as will conduct the current from the one electrode through the slag layer, which has become strongly heated in its vicinity and has thus become better conductive, to the metal, and from the lower end of the latter again through a slag layer to the other electrode.

From the character and simplicity of this method of working, without a doubt many results are to be expected.

Fig. 220 is from a photograph of the furnace with the electrodes in the position just given and ready for starting to work.

Thus the wishes were fulfilled with reference to the arrange-

FIG. 220.

ment of the electrodes which were expressed at the spring meeting of the German Electro-Chemical Society. It was further striven to secure for the furnace ready mobility of the electrode-holders and of the hearth, together with ample stability, and this may be said to have been achieved. As will be seen from the photograph, Fig. 220, the furnace has remained easily accessible from all sides. Nevertheless, when last winter, after three years had passed since the furnace was first used, I was again confronted with the problem of constructing experimental furnaces for the same working and current conditions, I felt compelled to decide once more in favour of *furnaces with independent electrodes* after the experiences hitherto gained. The number of problems which falls to the lot of a furnace constructed for all modes of heating, and with it the number of the positions of the electrodes, is far too large to be capable of solution with sufficiently simple electrode-holders arranged on the frames of the furnace. Even if a suitable construction could be found, it would again restrict the dimensions of the furnace in an undesirable manner, a defect which was several times felt with the furnaces just described. The arcs, fluxes, and other chemicals at hand for experiments give furnace charges and smelting products which deviate too widely from one another in conductivity, and also change too much in conductivity during the working, to permit, in the case of an experimental furnace, of the restriction in the dimensions of the smelting hearth which a fixed furnace frame would impose. During the four years the furnaces have been erected by me they were at first used very largely; later, however, almost solely for certain lecture experiments for which they were specially suited. The students mostly preferred to build up hearths themselves which were exactly adapted to their smelting experiments, by using a few bricks with suitably chosen linings, so as to be able to change the position of the electrodes at will during the experiment. Independent electrode-holders must, therefore, be at one's disposal in a testing laboratory; these and the necessary furnace-construction material will then suffice, provided one has an easily handled iron box at one's disposal for a few smelting operations which frequently come into question. The actual hearth-construction material, the hearth lining, mostly the charge itself, or part of it, should be capable of being stamped down into the box, so that the box will act as the support, and the cooling action of air or water can be made use of. The construction of the floor and the side walls of the hearth from fusible portions of the charge inside masonry has the disadvantage that they easily fuse with the rest of the charge, and contact with the masonry and contamination of the smelting products by it can now no longer be avoided.

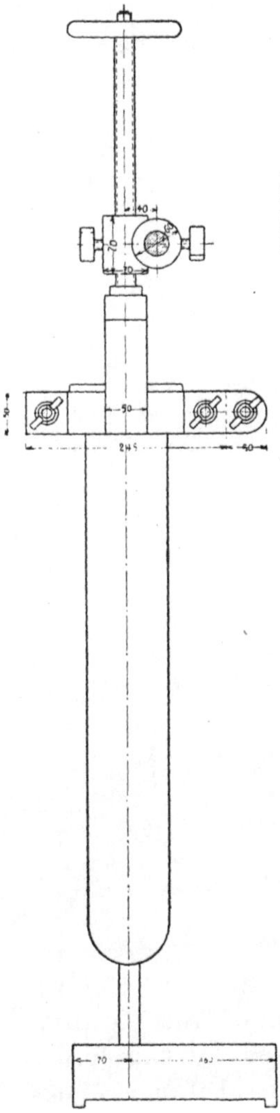

FIG. 221. Scale $\frac{1}{10}$.

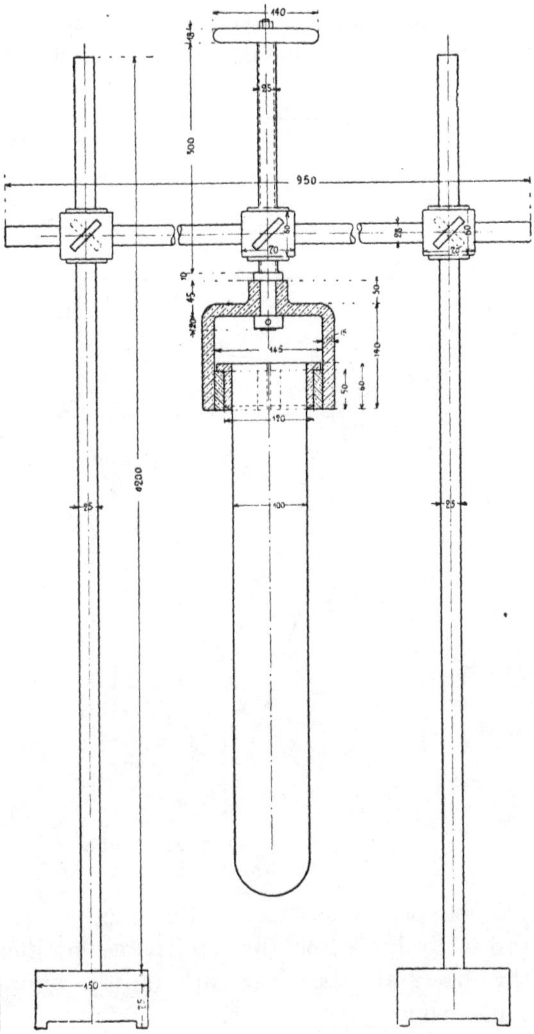

FIG. 222. Scale $\frac{1}{10}$.

For smaller furnaces (average currents of 300 amps.) the electrode-holders remain as shown in Figs. 211 to 215. For furnaces requiring currents above 300 to about 1000 amperes (for short tests somewhat higher currents can still be used) the actual electrode fittings, although made in the same way, are, of course, much larger in size, while each electrode has two stands, between which it is suspended from a cross-piece supported by the stands (Figs. 221 and 222). For the vertical

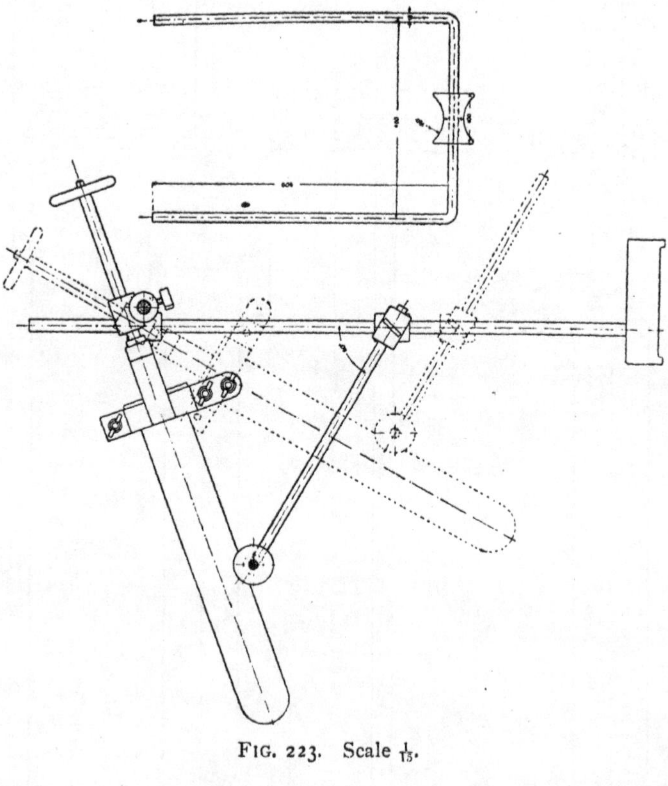

FIG. 223. Scale $\frac{1}{15}$.

position of the electrodes this method of attachment answers satisfactorily. For the horizontal position of the electrodes the necessary bearings are mostly afforded by the furnace walls, which render further supports or guides superfluous; for inclined positions these are, however, desirable, as the hearth walls cannot always be inclined as required by the electrodes, so that the latter have to stand free above the hearth wall. For this purpose a U-shaped rod provided with a roller is used, and is carried on the electrode uprights in

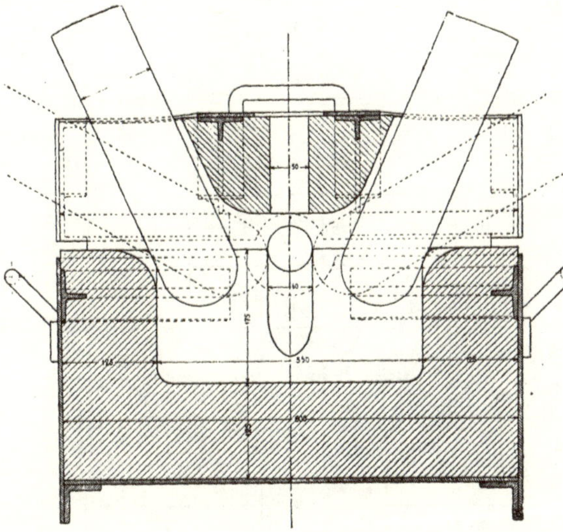

FIG. 224. Scales $\frac{1}{10}$.

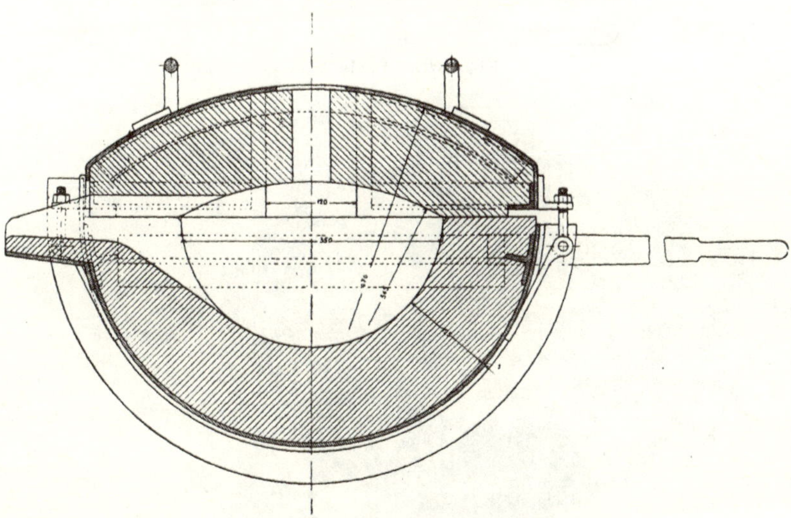

FIG. 225. Scale $\frac{1}{10}$.

sleeves. The foot of the stand must, in this case, be weighted, or bolted to existing foundation plates (Fig. 223).

For the construction of the furnace hearths either bricks are used with a lining which should be selected for each case, or iron boxes for frequently occurring work for which the hearth

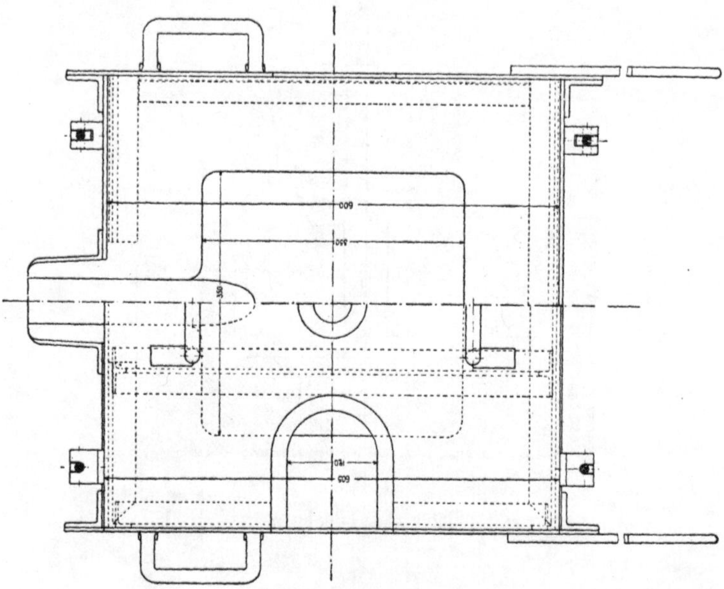

FIG. 226. Scale $\frac{1}{10}$.

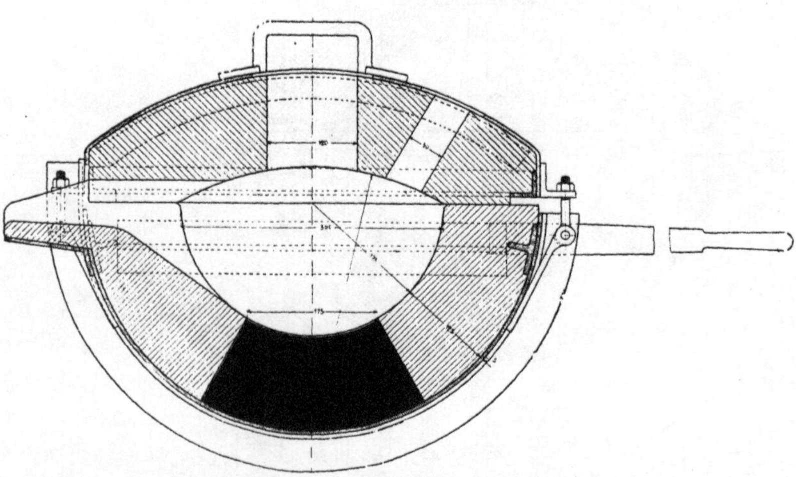

FIG. 227. Scale $\frac{1}{10}$.

dimensions are to some extent fixed. Of the latter I have shown two of the more suitable forms in Figs. 224 to 232.

Figs. 224 and 225 represent front and side sectional elevations of a semi-cylindrical hearth box with non-conducting lining, Fig. 226 showing the same furnace in plan.

When the smelting vessel and the products of reduction can

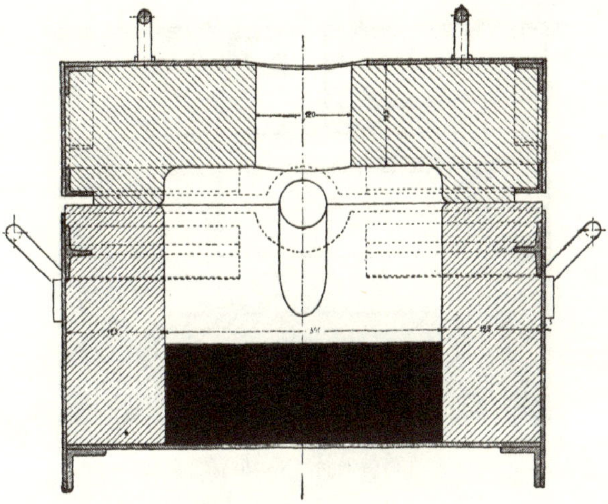

FIG. 228. Scale $\frac{1}{10}$.

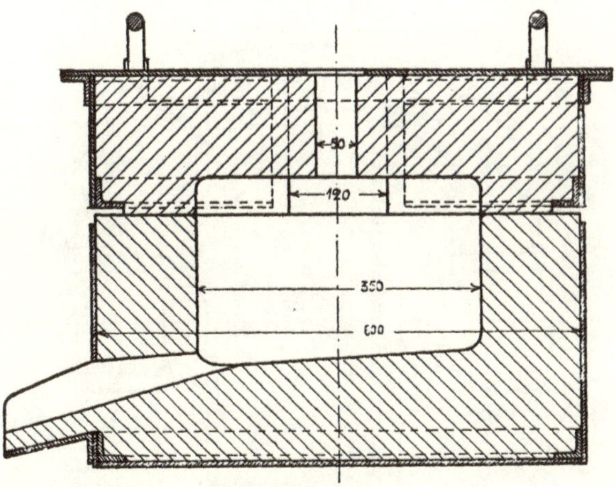

FIG. 229. Scale $\frac{1}{10}$.

be utilized as one of the electrodes, the floor should be partially lined with a carbon or metal block, as shown in Figs. 227 and 228 in side and front sectional elevation.

The semi-cylindrical shape of the furnace hearth, which is modelled on the Héroult steel furnace, enables the molten material to be easily poured out by tilting the hearth.

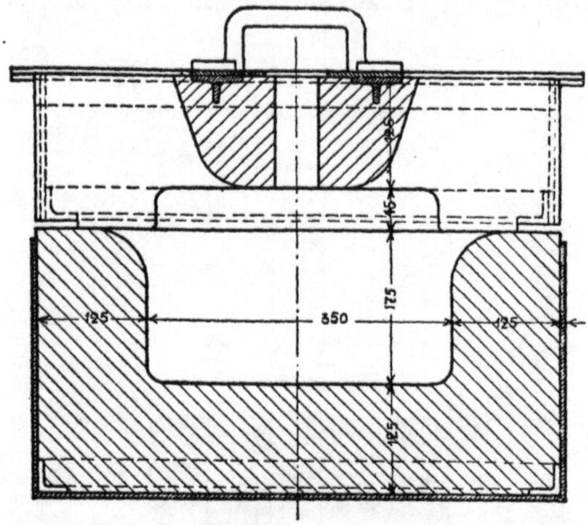

FIG. 230. Scale $\frac{1}{10}$.

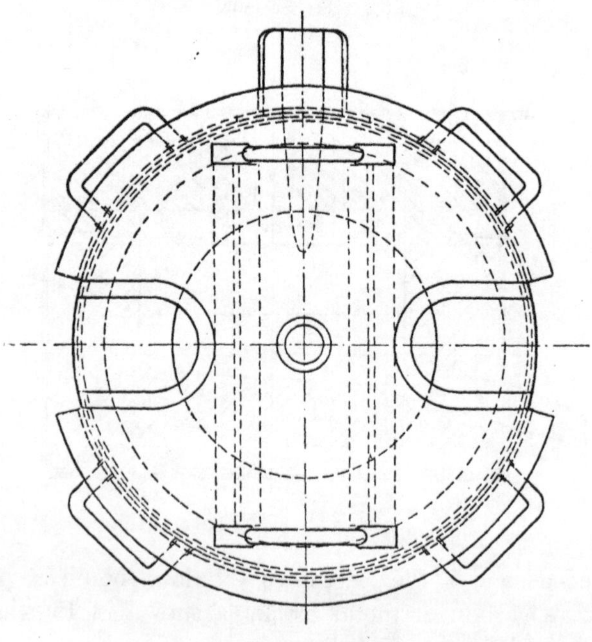

FIG. 231. Scale $\frac{1}{10}$.

For work with which it is preferable to tap the products of smelting, I have provided a low cylindrical hearth which can also be used for different modes of heating, and be lined in a manner

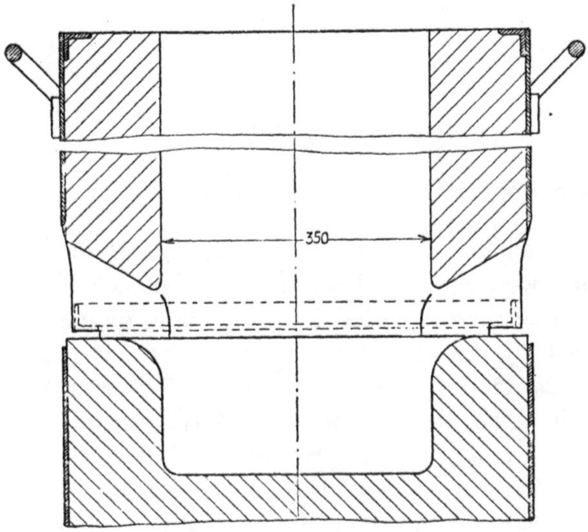

FIG. 232. Scale $\frac{1}{10}$.

similar to the hearths in Figs. 221 to 228. It is, however, better adapted to pit furnace working, for which purpose it would be provided with a suitable crown (Figs. 229 to 232).

The electrode holders and furnaces, Figs. 221 to 231, are constructed by Die Maschinenbauanstalt Humboldt, Kalk, near Köln.

CHAPTER VIII

THE CONSTRUCTION OF ELECTRIC FURNACES

General Remarks.—It was shortly after I started practice when the reports were circulated through the technical papers on the results, at that time regarded as astounding, of the experiments carried out by Ch. W. Siemens with his electrical smelting crucibles constructed during the years 1878 to 1880. As one can well imagine, these publications gave rise to vigorous expressions of opinion on the prospects of the practical application of the electric current for metal and other smelting processes. The views were mostly pessimistic. Doubt as to the practicability of constructing electric furnaces on a sufficiently large scale for industrial purposes, and doubt as to the possibility of generating electricity sufficiently cheaply for work of this kind, were combined with doubts as to the feasibility of procuring constructional material which would be able to withstand the temperatures of the electric arc (resistance heating at that time was not thought of at all). These doubts have fortunately been dispelled, and as regards constructional material we are better off than the ordinary furnace constructor, for with electric heating we can place the source of heat in the middle of the furnace charge. In the preceding chapters we have become acquainted with examples of electric furnaces which simply consisted of two electrodes and a piled-up heap of charge. One of the illustrations of these furnaces is again given here (Fig. 233). Thus, *the primary materials of construction for the electric furnace are the electrodes and the charge.* This, speaking figuratively of the other metallurgical

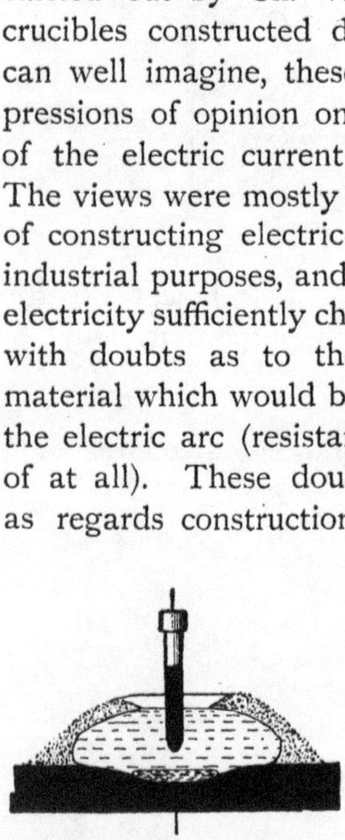

FIG. 233.

furnaces, forms the inner lining of the furnace. If masonry be used at all, it only takes the part of the outer shell of metallurgical furnaces, *i.e.* it forms the heat protection or support of the inner lining. In many cases the outer shell, or masonry, is not required at all; the charge, or a portion of it, is pounded down, or melted, into metal vessels, metal shells, which under certain circumstances act as one of the electrodes, or supply connections. By erecting these vessels so that they are freely exposed to the air, the cooling action of the air can be utilized to obtain a solidified layer of the charge. Where this does not suffice we can, of course, use water-jackets, hollow-walled metal vessels, pipes, and other hollow metal bodies for the same purpose. As the most essential *elements in furnace construction* we must, therefore, consider the electrodes, their *connections* and *holders*, and the *charge* itself with reference to its introduction and arrangement as far as it enters into consideration as the furnace wall, resistance, or intermediate electrode.

ELECTRODES AND TERMINAL CONNECTIONS

Electrodes.—In electric furnaces not only are those carbon, or metal, bodies designated electrodes which act in the accepted sense as electrodes, but also the end connections, mostly consisting of carbon, more rarely of cooled metal bodies, which connect the heating resistances to the supply circuit, and should therefore be more correctly termed resistance connections. As, however, with electrolytically conducting resistances these naturally play the part of the electrodes, the name "electrode" may also be kept for the actual resistance connections.

For electric furnaces carbon constitutes, for reasons which can be readily understood, the most valued of the materials for the electrodes. We also know that the amorphous varieties of carbon—charcoal, coke, soot—become all the more conductive the more strongly they are heated, I should say, the more nearly they are brought by heating alone or by the agency of carbide- and graphite-forming impurities and intentionally added ingredients (cp. pp. 66 to 70) to the graphitic state in which carbon, as is well known, conducts the best. Of all metallically conducting substances carbon withstands the highest temperatures. In the temperature of the electric furnace, however, it is always highly reactive, on the one hand acting as a powerful reducing agent, and on the other easily dissolving in all metals which readily form carbides. If these actions are to be avoided, the carbon electrodes must be cooled, or electrodes should be used composed of the metals to be melted, and in this case they also must be cooled to prevent them from dissolving. Finally, in the cases considered in the preceding chapters the means have been adopted which were proposed by *Héroult*, *Ferranti*, and *Kjellin* for the avoidance of immediate contact of the electrodes with the product of the smelting, or for working entirely without electrodes. However, all things considered, the universal practice is to use carbon as the material for the electrode in the construction of electric furnaces. Metals are only then considered for the electrodes, even so mostly in a secondary degree, when they are to be melted either electrolytically or electro-thermically as products of smelting. Even in the electrolytic deposition of the metal in the electric furnace the opposite pole consists almost without exception of carbon.

I am unable here to enter into the details of manufacture of carbon electrodes, although in the larger factories using electric furnaces the manufacture of electrodes mostly forms a portion of the total work. Excellent information on the manufacture and testing of electrodes is given in Dr. Zellner's book [1] which appeared a few years ago. We can only recommend every one using electrodes to at least study their manufacture, especially in connection with the method of shaping them, so as to avoid requirements as to electrode forms which either cannot be met, or can only be carried out at heavy costs to the user's detriment. Massive or hollow rods or plates having a uniform cross-section throughout their length are the least difficult, and therefore the least costly, to manufacture. When the cross-sections are not very large, the carbons are made in presses from which the material is forced under pressure in an endless band (rod) to be

[1] Dr. J. Zellner, "Die künstlichen Kohlen für elektrotechnische und elektrochemische Zwecke, ihre Herstellung und Prüfung." Berlin: J. Springer. 1903. [The following may also prove useful: "The Manufacture of Carbons for Electric Lighting and Other Purposes," by Francis Jehl. London: "The Electrician' Printing and Publishing Co., Ltd.]

cut up into the desired lengths. More massive blocks are formed in presses with core-boxes and rams, and here also the forms should be as simple and regular as possible to ensure uniform pressure in pressing and to facilitate the removal of the compressed body from the mould. When irregularly shaped carbon bodies are unavoidable, they should be built up of simpler moulded forms.

From communications which I have received from the larger carbon factories,[1] cylindrical or prismatic bars of square or rectangular cross-section can at any time be made with existing moulds in any desired sizes not exceeding 100 mm. (4 in. nearly) in diameter; for diameters above 100 mm. the sizes increase in steps of 50 mm. (2 in. app.) up to diameters of 400 mm. (15¾ in.), and up to 1600 mm. (5 ft. 3 in.) in length.

According to the usual or desired method of manufacture of the electrode connections, the extremities of the electrodes themselves are correspondingly bored out, hollowed, or cut out, or are provided with other suitable changes in form which are either made during the moulding process or are subsequently produced.

Terminal Connections.—The simplest provision for the connection pieces consists in making longitudinal or transverse holes in the electrodes. These holes are mostly for receiving or passing through the screw bolts; the longitudinal holes, however, being for the reception of contact pieces. For the latter purpose the holes need not extend throughout the length of the carbon electrode, but only to a moderate depth (100 to 200 mm.) from one end. It should be borne in mind that metals—and these are always used as the terminal connections for the carbons—expand more than carbon; clamping bands, plates, nuts, and such metal bodies which fit on the electrode ends like a coupling or cap naturally fit looser when hot than when cold. A bar or rod which is let into the carbon, on the other hand, makes more intimate contact with the carbon on being heated. Such contact rods should, however, only be of moderate dimensions to prevent the carbon-head from springing.

[1] C. Conradty, Nürnberg; Dr. Albert Lessing, Nürnberg; "Planiawerke, Aktiengesellschaft für Kohlenfabrikation," Berlin, N.W., 7, Dorotheenstr. 45; Schiff & Co., Schwechat bei Wien.

A few of the methods of fastening metal rods in the holes made in the carbon electrodes are shown in Figs. 234 to 239.

In connection with Fig. 235 it should be pointed out that the metal rod which widens at its lower end is surrounded by metal plates which are inserted when the rod is in its lowest position in the bore of the carbon; they are drawn up with the rod when it is raised.

In some of the illustrations of the furnaces given in the pre-

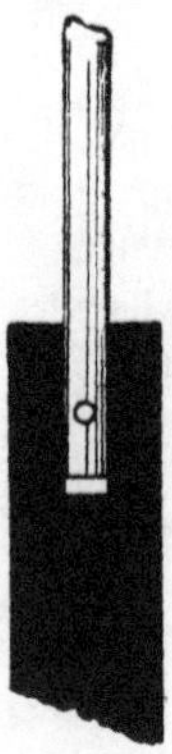

FIG. 234.

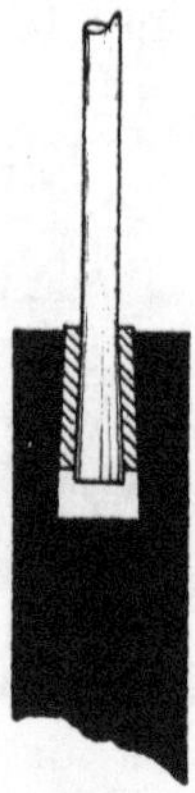

FIG. 235.

FIG. 236

FIG. 237.

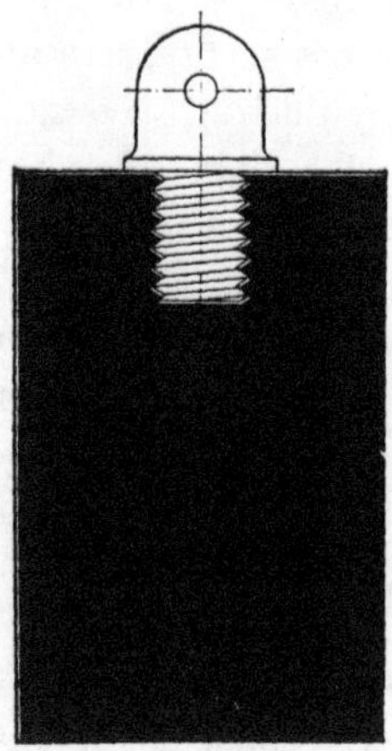

FIG. 238.

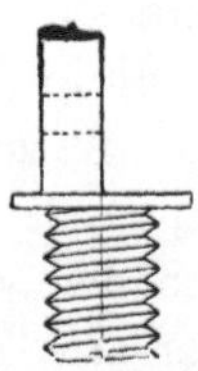

FIG. 239.

ceding chapters screw threads will be seen indicated on the ends of the metal rods let into the carbons. Coarse threads, and such suffice, can be cut in the carbons. It might, however, also be possible to compress and bake with the carbon blocks short screw bolts, nuts, and small blocks with screws similar to foundation bolts. Made from cast-iron such attachments are very cheaply prepared, withstand the temperatures of the baking ovens, are thus intimately connected to the carbon, and not only allow the conductors to be screwed on, but the carbon rods to be jointed together to prevent waste of the carbon ends. This method of jointing together pieces of the electrodes is naturally only permissible when metals, alloys, or compounds are to be fused in which the iron connecting pieces, which finally become melted with the consumption of the electrode, do not produce any contamination.

When the electrodes have to be introduced through thick furnace walls or into deep furnace hearths or pits, in which very high temperatures are maintained, and in consequence of which one cannot venture too far forward with the metal electrode holders which are always used, one has to put up with comparatively large quantities of electrode ends.

When the smelting hearths are lined with carbon, this, of course, affords an opportunity of making use of the carbon remnants. The consumption of carbon for linings does not, however, always keep pace with the supply of the electrode remnants, so that in these cases too it is important to keep down the quantity of the latter.

Carbon Electrodes of Urbanitzky and Fellner.—Already in their first patents for an electric pit furnace Urbanitzky and Fellner[1] recommended carbon anodes with dovetailed, tenoned and mortised ends, so that the carbon plates could be pushed forward in a continuous manner (Fig. 240).

FIG. 240.

Others have even recommended attachments with screw threads at the one end of the electrode and corresponding tapped holes at the other for piecing together the carbon rods.

Such requirements very greatly increase the difficulty of com-

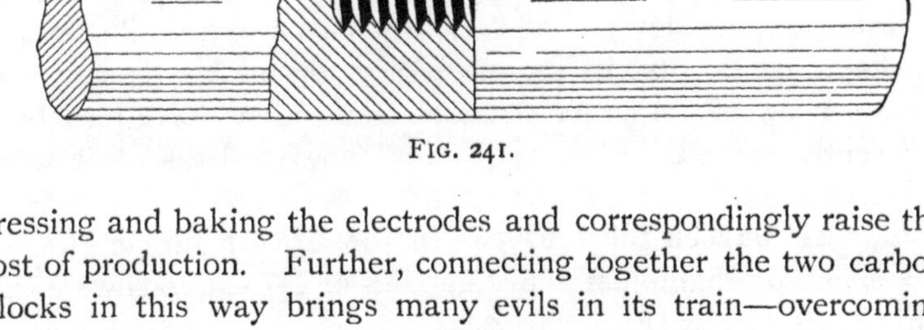

FIG. 241.

pressing and baking the electrodes and correspondingly raise the cost of production. Further, connecting together the two carbon blocks in this way brings many evils in its train—overcoming

[1] English Patent, No. 6965 of 1893.

the contact resistance at the junction incurs loss of power; the joint easily works loose and intensifies the first fault; or the carbon end in the furnace falls off, producing short-circuits or other disturbances; when the joints are firm the danger of rupture of the joined carbon bodies is again large, corresponding to the pressures which occur here.

It is in most cases more correct to arrange the electrode-holders (compare these) so that the electrodes are consumed until as little remains as possible. In works in which the consumption of electrodes forms a considerable item in the manufacturing cost the electrodes will always be manufactured at the works, and the remnants suitably utilized.

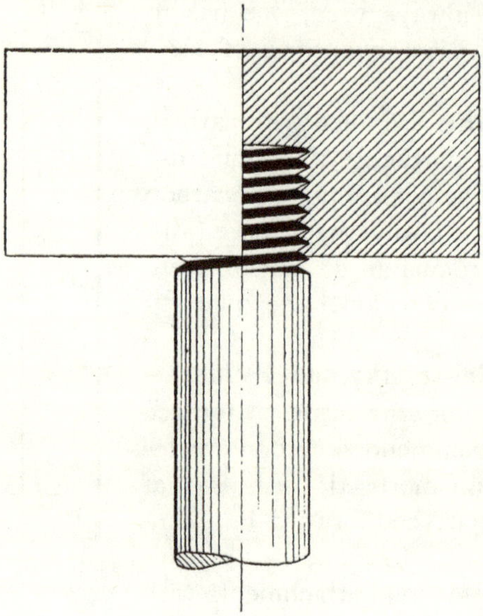

FIG. 242.

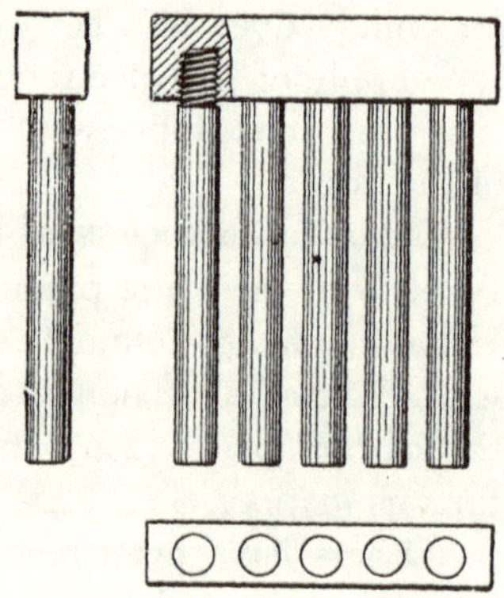

FIG. 243.

The manner of fixing the small electrodes in copper, bronze, or iron clamping devices is so clearly shown in the illustrations of the furnaces that it is unnecessary for me to repeat them here. A few of the larger electrode connections have also been sufficiently noticed. Figs. 241 to 243 show methods of securing the electrodes.

Cowles' Carbon Electrodes.—In the Cowles furnace for the production of aluminium alloys bundles of carbon rods were cast into bronze blocks (Figs. 244, 245).

Acheson's Carbon Electrodes.—Acheson places outwardly

projecting copper strips between the carbon blocks, by means of which he makes connection to the resistance cores. The whole

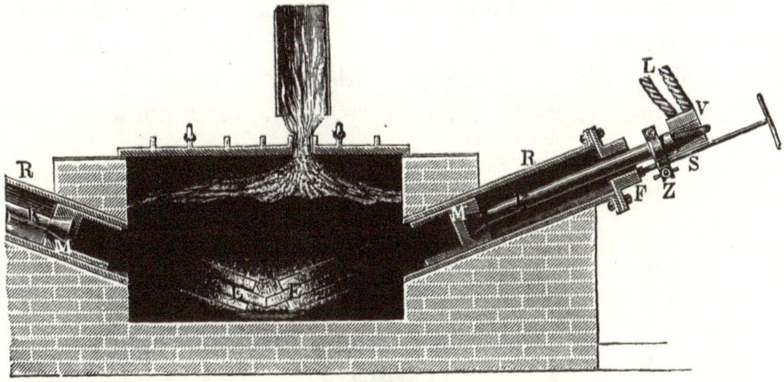

FIG. 244.

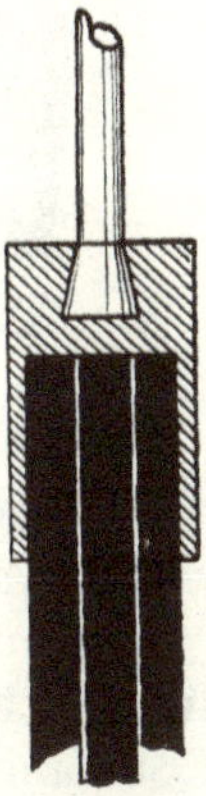

FIG. 245.

bundle of carbons is then clamped in an iron frame, and, finally, the strips are bent together and secured to the supply cables (Fig. 246).

Carbon Electrodes of the Willson Aluminium Company.—Electrodes which widen at their ends like a trapezium where the junction is made are not the simplest to press. They certainly afford a very convenient contact, however (Fig. 247). They were used in the first furnaces of the Willson Aluminium Company.

The carbon factories mentioned above also furnish their larger carbon blocks with tapered contact surfaces in the form of dovetailed ends (Fig. 248).

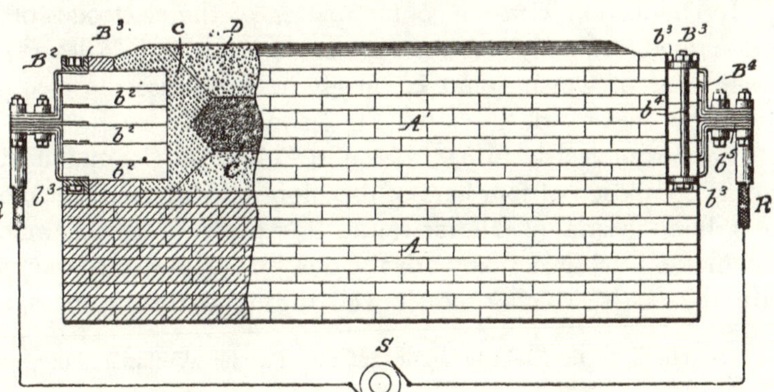

FIG. 246.

Such tapered surfaces need not be made in the press, or cut, provided the space in the furnace will permit of the use of wide electrode heads. By putting on metal plates tapering on the outside the same kind of connection will be obtained. The contact plates will be firmly pressed against the carbon by the cheeks of the electrode holder (Fig. 249). The connection can

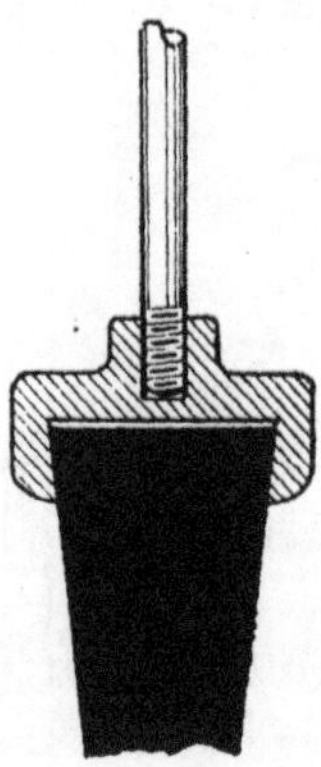

FIG. 247.

FIG. 248.

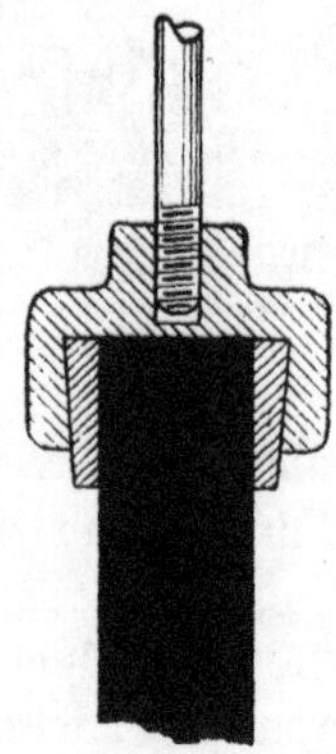

FIG. 249

be still further secured by the insertion of small pins through the wedges and carbon block.

Terminal Connections of the Société des Carbures Métalliques and of Keller, Leleux, & Co.—The furnaces of these two Companies shown in actual operation at the Paris Exposition were furnished with very good connections to the electrodes.

In connection with the contact plates of the electrodes of the first-named company screw bolts were converted into levers, the lower arms of which could be firmly held against the electrode heads by the nuts of the bolts (Figs. 250, 251).

For large electric furnaces Keller, Leleux, & Co. manufacture blocks composed of four bars of the highest conductivity enclosed in carbon material of inferior conducting power.[1] In this manner the makers achieve many advantages. It is easier to manufacture thin bars, which conduct well throughout their mass, than

[1] Borchers, "Die Elektrochemie auf der Pariser Weltanstellung," 1900. W. Knappe, Halle a. S.

thick, massive blocks. And by packing these bars in inferior material the more valuable material is protected from the unavoidable loss arising from overheating and burning. Another feature of very special importance is the complete utilization of

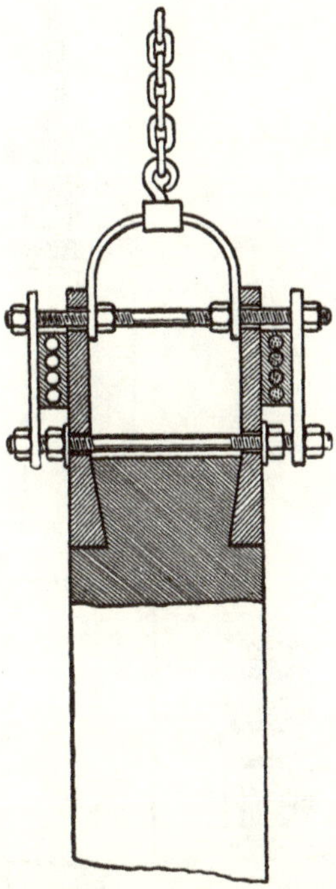

FIG. 250,

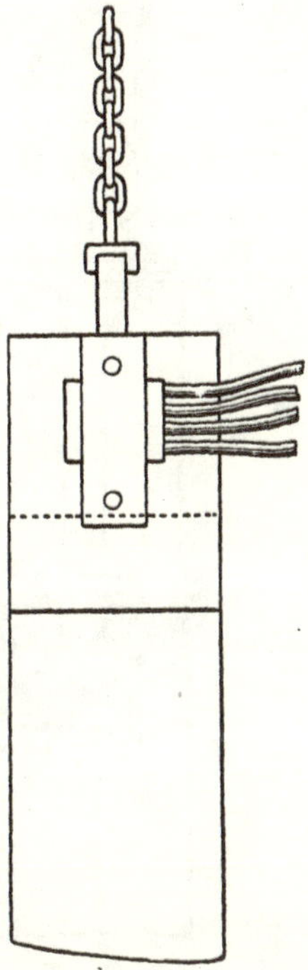

FIG. 251.

the short ends of the electrodes which after some time remain over in the holders. This is effected by pulverizing the carbon ends, mixing the powder with a binding material, and cementing together the four good carbon bars with this mass. The bundle of bars made up in this way has then to be consolidated by baking in a muffle furnace.

Copper strips are firmly clamped on the ends of the four centre bars projecting beyond the carbon block at the top by means of iron plates and bolts which terminate in wedge-shaped blocks. The other ends of the copper strips are clamped together with the ends of the cables by means of iron plates (Figs. 252, 253, 254).

Heroult Carbon Electrode Holders.—In the new Héroult furnaces four iron plates are placed round the top ends of the electrodes. The one plate is fastened to the horizontally projecting arm of the electrode carrier, and the other three, to

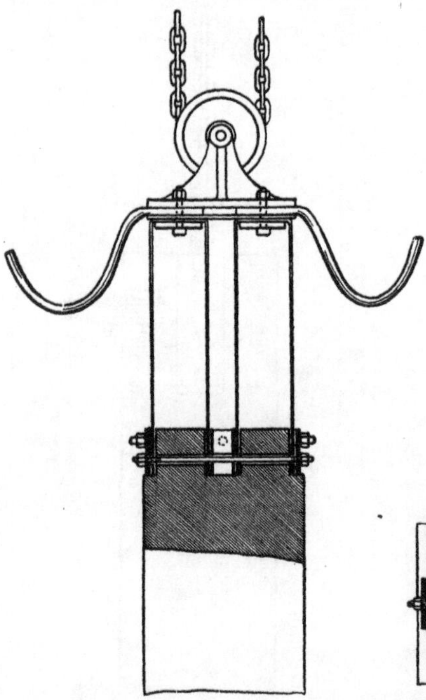

FIG. 252.

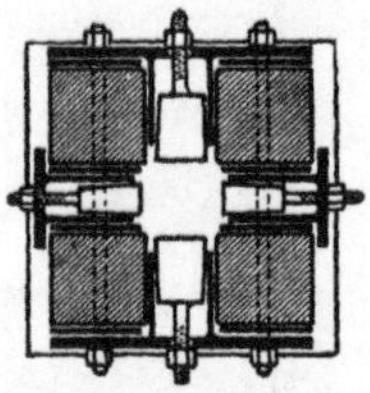

FIG. 253.

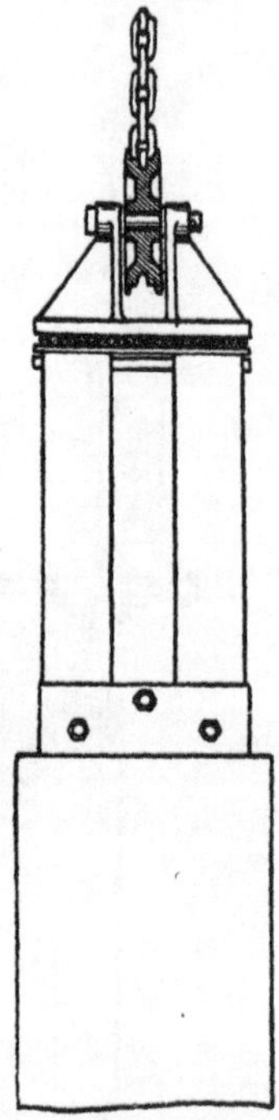

FIG. 254.

which the main cables are screwed, are firmly held against the electrode by flexible metal bands, which can be tightened by a screw working in the carrier arm. In this way the electrode bears firmly against the first plate (Figs. 255, 256, and 257).

Mayer's Carbon Slab Electrode.—For closing the floor of an electric furnace Mayer [1] uses an electrode slab, consisting of carbon encased in a metal, on the underneath side of which he provides projections with conical holes. The copper heads of a lever mechanism engage in these holes, and when the lever is drawn back the plate rests on the supporting ledges in the side walls of the furnace (Figs. 258, 259).

[1] German Patent, No. 101,131 of May 29, 1898.

Various constructions of cooled holders for small electrodes have already been given in connection with the experimental furnaces (Figs. 126–130).

Schindler's [1] Cooled Electrode Holders and Terminal Connections are made on the pattern of the *Lürmann* slag moulds

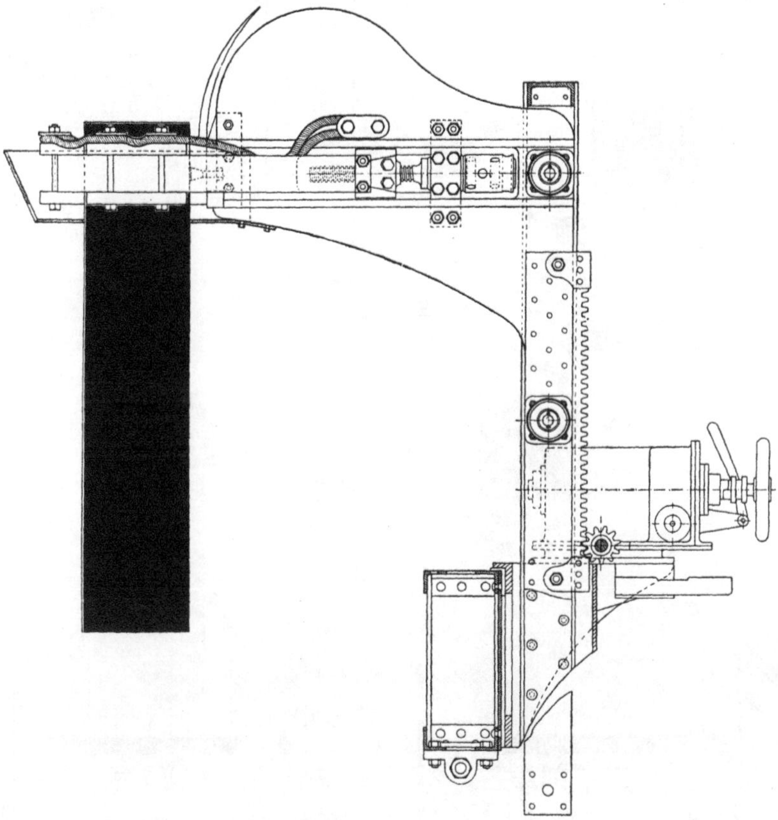

FIG. 255. Scale $\frac{1}{25}$.

by casting suitable piping into larger metal bodies (Figs. 260–264).

Cooling Devices of Urbanitzky.—The cover of Urbanitzky's [2] furnace, besides the chamber for the cooling water, contains four

[1] United States Patent, No. 573,041 of December 15, 1896.

[2] German Patent, No. 82,164 of January 29, 1895; English Patent, No. 7265 of 1895.

water-cooled electrode guides, and a similarly cooled feed pipe. The pit wall is closed at the top by means of an annular cooling ring. In order during the working to be able to rotate the cover

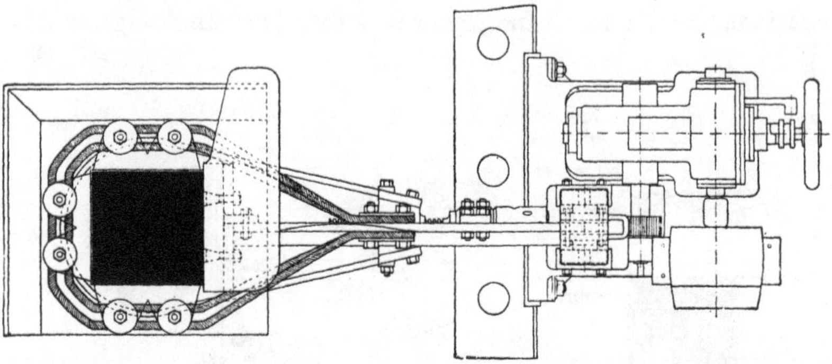

FIG. 256.

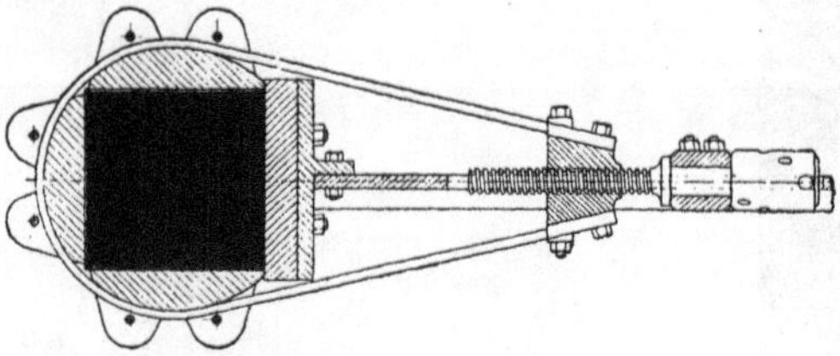

FIG. 257.

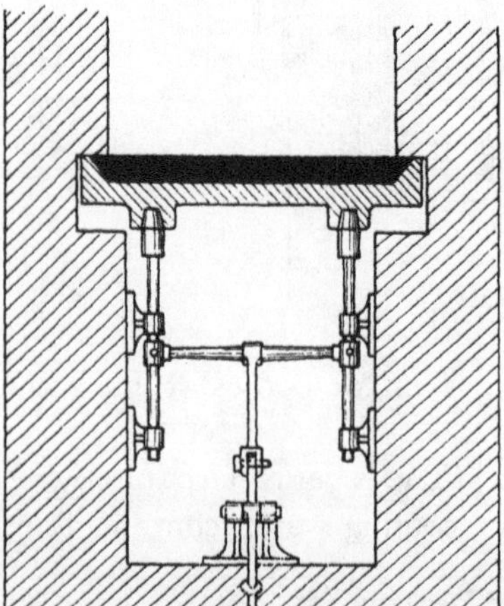

FIG. 258.

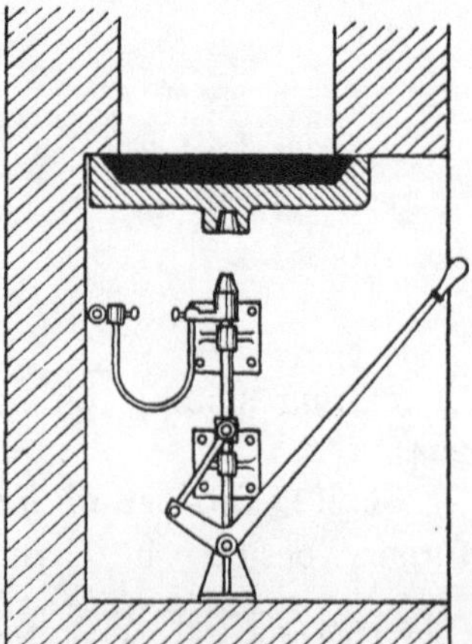

FIG. 259.

with the electrodes, the underneath side of the cover and the upper side of the ring are provided with annular grooves in which are the balls which support the cover (Fig. 265).

Cooling Devices of de Chalmot.—In an American patent specification de Chalmot[1] also gives various methods of constructing cooled covers for small and large electric furnaces. A few of these arrangements are depicted in Figs. 266–269.

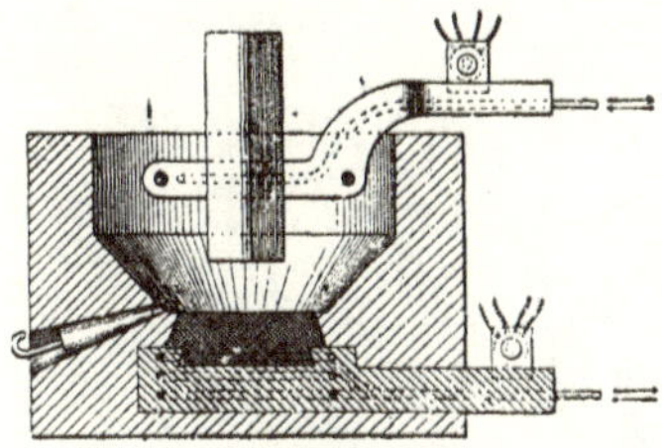

FIG. 260.

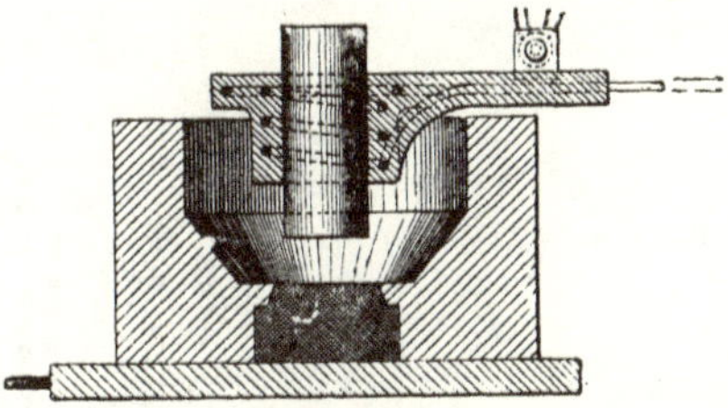

FIG. 262.

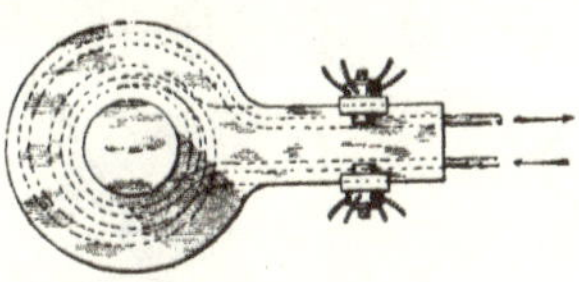

FIG. 263.

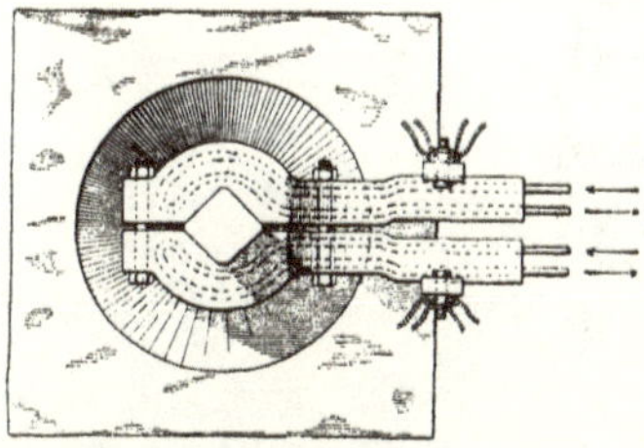

FIG. 261.

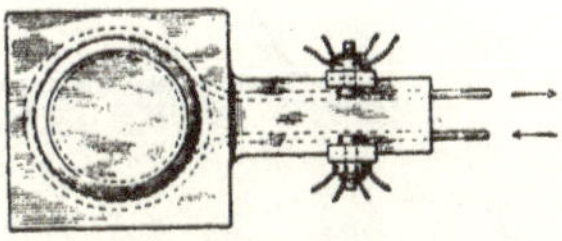

FIG. 264.

Cooling Device of Urbanitzky and Fellner.—In the Urbanitzky and Fellner furnace[2] one of the electrodes is arranged to be air-cooled. The furnace hearth has an inverted metal pan let into it, the cover of which is provided with air inlet-holes. The air is sucked out through the pipes projecting into the pan, and an air-chest attached to the cover (Fig. 270).

Patin's Cooled Electrode Holders.—In the cooled electrode holders of Patin the electrode is not inserted direct, but is clamped

[1] United States Patent, No. 699,654 of May 13, 1902.

[2] German Patent, No. 82,164 of January 29, 1895; English Patent, No. 7265 of 1895.

by a flexible split bush in the holder, which slightly narrows like a cone (Figs. 271 and 272).

Morehead's Electrode Holder.—For the Horry furnaces Morehead[1] has combined very efficient electrode holders with a

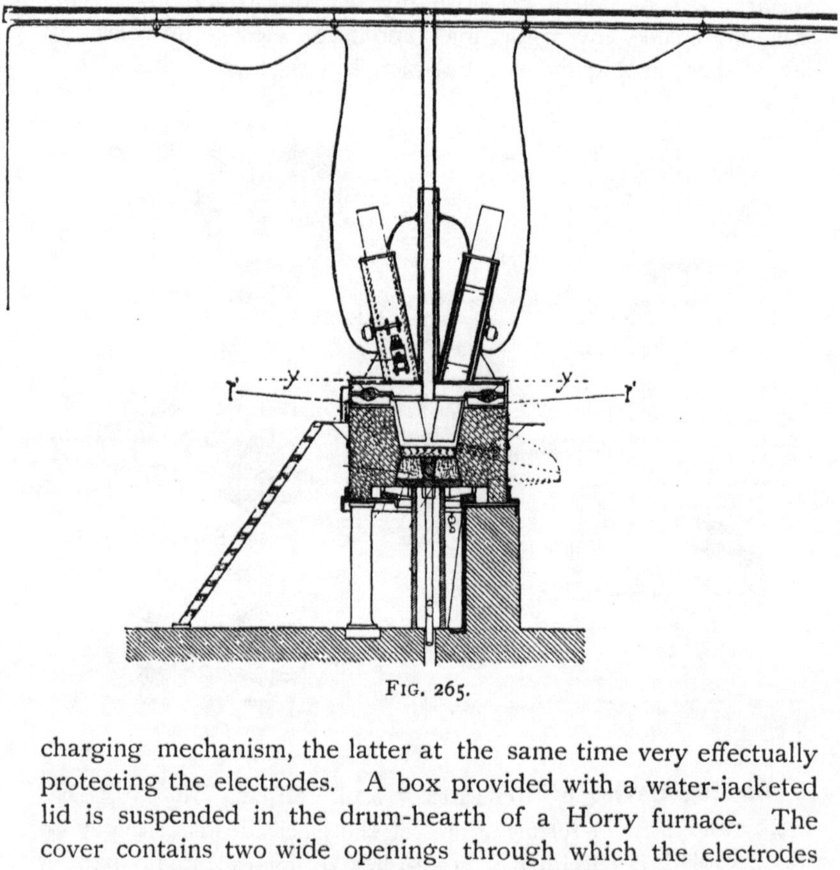

FIG. 265.

charging mechanism, the latter at the same time very effectually protecting the electrodes. A box provided with a water-jacketed lid is suspended in the drum-hearth of a Horry furnace. The cover contains two wide openings through which the electrodes are inserted in insulating bushes. In addition it carries two wide pipes, of which the one is for the introduction of the charge, while the other leads off the gases due to the reaction ; and there are also two narrower inlet and outlet tubes for the cooling water of the cover (Figs. 273 and 274).

Electrode Holders of Price, Cox, and Marshall.—The holders designed by Price, Cox, and Marshall with protecting jackets for

[1] United States Patent, No. 782,917 of February 21, 1905.

the electrodes, were also principally intended for Horry furnaces. The electrodes are composed of graphite rods, of which several are secured to a common head (Figs. 275–278). This head 1 comprises a cast-iron hollow box, with a lug 2 for connection to the mains. The hollow space 3 can be water-cooled by means of

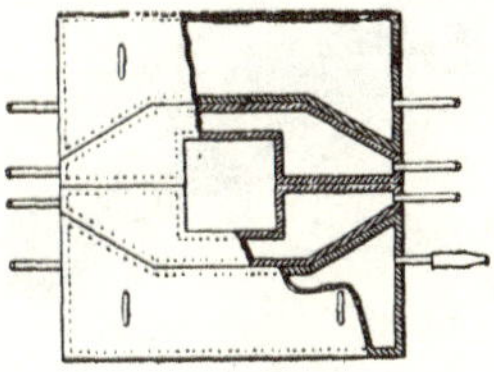

FIG. 266.

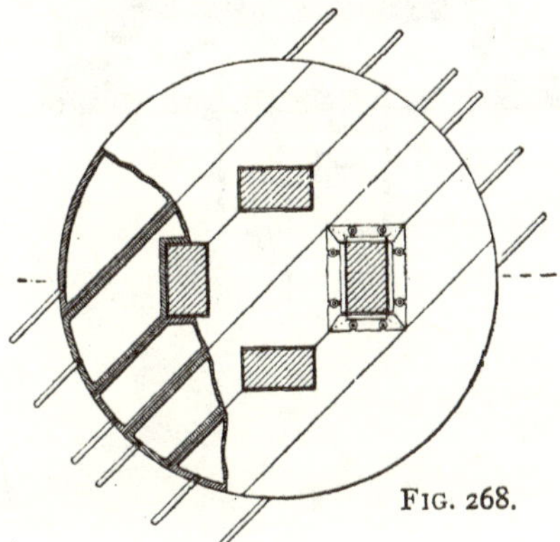

FIG. 268.

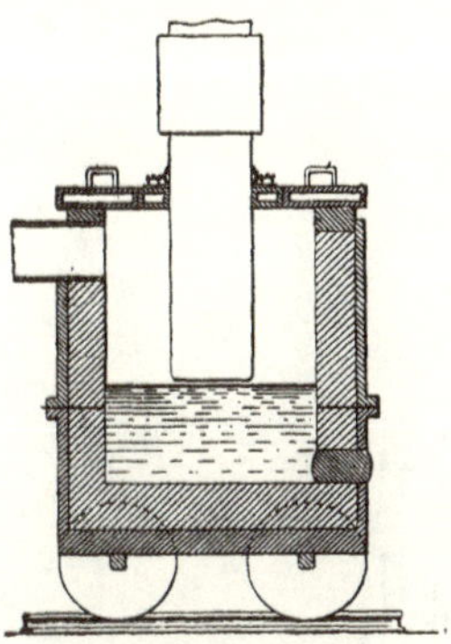

FIG. 267.

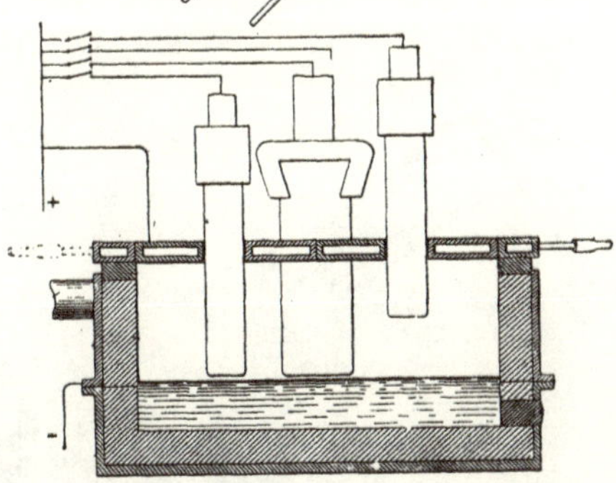

FIG. 269.

the pipes 4 and 5, the course of the water being controlled by a partition, 6. The actual electrode holders, which are composed of the nipples 8 and caps 9, are screwed into a series of tapped holes 7 in the floor of the hollow box in such a manner that the caps 9 are cooled by the circulating water. The graphite rods 10 are screw-threaded and are screwed into these caps. Electrodes built up in this manner, and suitably insulated, are also inserted in the lids of electric furnaces as shown in Figs. 275–278. This arrangement is the one specially used in the well-known Horry furnaces. In this case the main charging mechanism and gas outlet pipes are also placed in the cover. To protect the graphite rods from the air, they are enveloped by a sheet-iron or wire-gauze covering

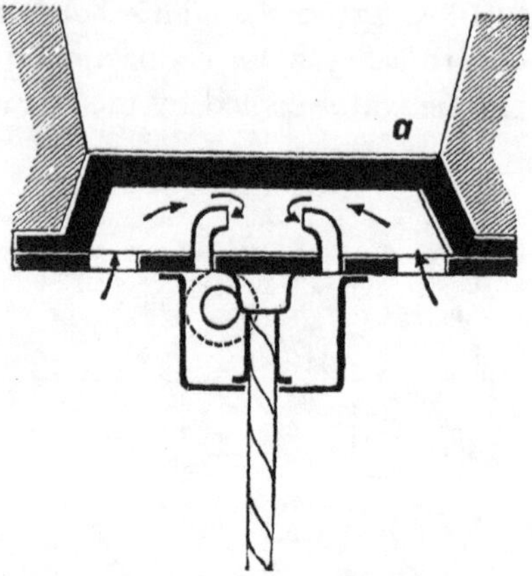

FIG. 270.

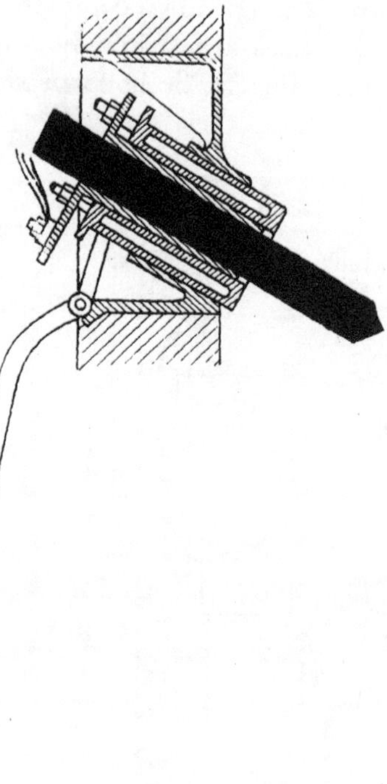

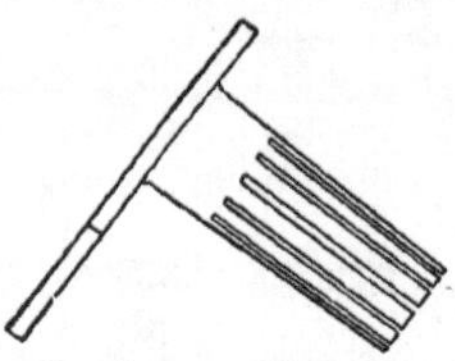

FIG. 272.

FIG. 271.

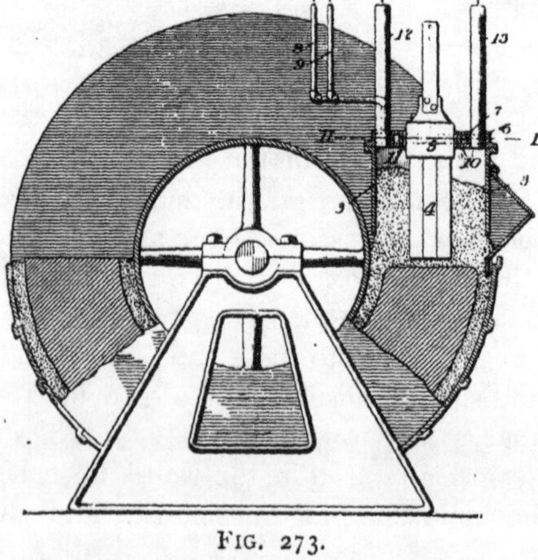

FIG. 273.

after the rods have been first smeared over with tar or a mixture of tar and breeze. Other protecting materials are recommended,

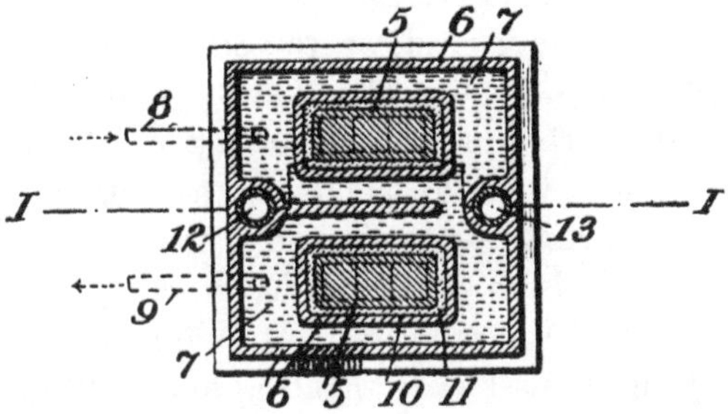

FIG. 274.

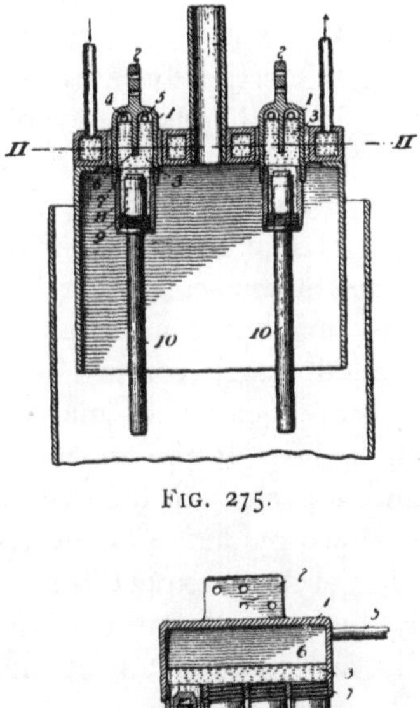

FIG. 275.

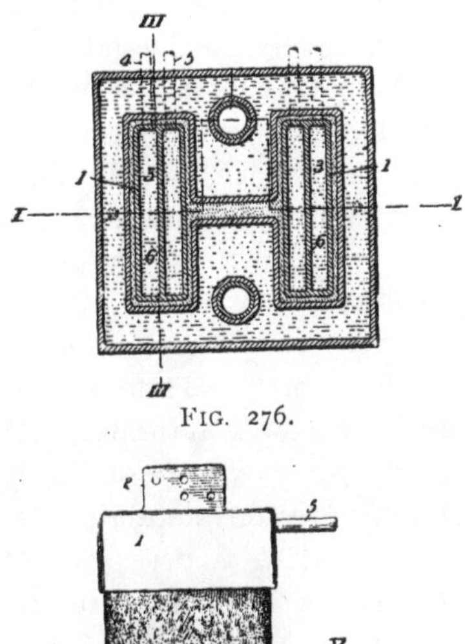

FIG. 276.

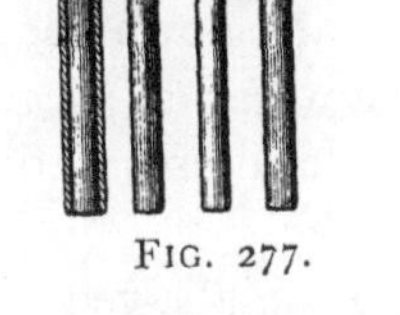

FIG. 277.

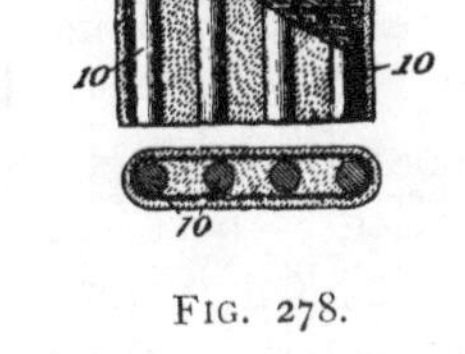

FIG. 278.

such as asbestos, or clay and coal-dust, also admixed with siloxicon, but always with a gauze envelope to avoid crumbling.

The inventors state that the heavier cost of the graphite rods is equalized by smaller thickness and smaller length; also that by means of their protecting devices just mentioned the rods last a long time and can be used down to very short ends.

Dimensions of Carbon Electrodes. — In determining the dimensions of the carbon electrodes, we have first of all to decide whether or not we wish to avoid all heating of the carbons conducting the current to the furnaces that can be practically prevented. The decision in the affirmative appears to be the natural one, although the other forms by no means a rare exception. Already in discussing arc heating and combined resistance and arc heating I pointed out that, for the production of the highest degrees of heat, making the cross-sectional area of the electrodes amply large to carry the current was not to be recommended. Carbons which conduct well electrically are also comparatively good conductors of heat. It is, therefore, possible for them to conduct so much heat away from the furnace to the outside that the desired maximum temperature is confined to a very small space, namely, the narrowest arc zone, whereas outside this everything solidifies. This defect can, of course, be remedied to a certain extent by making the heating space as small as possible, and by enclosing it in thick-walled vessels comprised of bad heat conductors, which, however, store heat well (magnesite, lime, etc.). Time is then required to heat up the apparatus until it becomes saturated with heat, and, especially in the case of testing work, or in periodically melting down moderate quantities of difficultly fusible materials, often a larger heat expenditure is involved than is necessary in quickly fusing with carbons, the dimensions of which are intentionally made too small, at any rate at the extremities of the electrodes inside the furnace. For rapid tests of the behaviour of a substance at the highest temperatures, overloading the electrodes offers a very convenient expedient. Rules cannot, however, be laid down for abnormal working conditions of this character. How far an electrode may be overloaded is best seen from a few figures relating to the permissible load that can be used.

The following are the cross-sections to be allowed per ampere for different sizes of carbon rods :—

Diameter of carbon up to		Cross-section of carbon per ampere.	
Mm.	In. (app.)	Sq. mm.	Sq. in.
50	2	10	0·0155
100	4	12	0·0186
200	8	20	0·0310
300	12	30-40	0·0465-0·0620
400	16	60-90	0·0930-0·1395

The reason for the remarkable increase of the cross-sections per unit of current in the thicker carbons is on account of the decreased homogeneity with increasing section. The thicker the rods the less uniformly they glow even with uniform furnace temperatures and for uniform intervals of time. The binding materials used for the carbon do not always undergo regular changes, and their volatile products do not always escape by the same regular paths. It is, therefore, advisable with very large electrodes to build them up of smaller, better conducting carbon rods. This course was, in fact, always pursued in the early days of the development of applied electro-chemistry, and has recently been again followed with considerable success by Keller and Leleux in the characteristic manner described on p. 190. For furnaces working with currents exceeding 1000 amperes, it is better to employ an electrode with special current leads for every 1000 amperes than to send thousands of amperes through thick, badly conducting blocks, the handling of which in addition gives rise to great inconvenience on account of their volume and weight. In the case of pure electro-thermal processes for very high temperatures, considerably more than 1000 amperes can, of course, be carried per electrode, for the reasons stated above.

Weight of Carbons.—Although the furnace constructor should know the weight of the carbon electrodes in calculating the dimensions of the electrode holders for reasons of strength, in the majority of cases it will be found that the metal rods, plates, and cables, which serve both as the current leads and electrode holders, will have the requisite strength for supporting the electrodes when their sections have been properly dimensioned for the current. The electrodes are, however, sold by weight, and in determining the working costs the consumption of the electrodes must be calculated by weight. As a guide it may be of use to state that good conducting, compactly baked carbon blocks weigh very nearly 600 kg. per cubic metre (6 gm. per cubic centimetre = 375 lb. per cubic foot app.).

Electrode Fittings (Connections).—The electrode fittings (connections), of which a few of the simpler forms have just been discussed with the electrodes themselves, mostly consist of iron. Iron is also largely used for the holders and guide rods from which the electrodes are generally suspended, and which no doubt carry the currents when these are not very heavy. In the latter case copper conductors would be mostly connected to them.

I shall only give the smallest permissible cross-sections of copper to be used with increasing currents. For iron the values are to be multiplied by 6.

Smallest Permissible Cross-Sections of Copper with Increasing Currents

Current in amperes.	Diameter.		Cross-section.		Voltage drop.	
	Mm.	In.	Sq. mm.	Sq. in.	Volts per km.	Volts per 1000 yd.
1	0·38	0·015	0·113	0·00017	165	151
5	1·09	0 043	0·93	0 00144	100	91
10	1·75	0·069	2·40	0·00295	77·7	71
15	2·29	0·090	4·12	0·00639	68·1	62·3
20	2·77	0·109	6·03	0·00935	62·1	56·8
25	3·20	0·126	8·04	0·01245	58·1	53·1
30	3·61	0·142	10·2	0·01581	54·8	51·0
35	4·01	0·158	12·6	0·01953	51·8	47·4
40	4·37	0·172	15·0	0·02325	49·9	45·6
45	4·72	0·186	17·5	0·02712	48·1	44·0
50	5·08	0·200	20·3	0·03146	46·1	42·2
55	5·41	0·213	23·0	0·03565	44·7	40·9
60	5·72	0·225	25·7	0·03983	43·7	40·0
65	6·05	0·238	28·7	0·04449	42·3	38·7
70	6·35	0·250	31·7	0·04914	41·3	37·8
75	6·55	0·258	33·7	0·05223	40·3	36·8
80	6·96	0·274	38·0	0·05890	39·3	35·9
85	7·24	0·285	41·2	0·06535	38·6	35·3
90	7·52	0·296	44·4	0·06882	37·9	34·6
95	7·80	0·307	47·8	0·07409	37·2	34·0
100	8·08	0·318	51·3	0·07951	36·5	33·4
110	8·61	0·331	58·2	0·09021	35·3	32·3
120	9·09	0·358	64·9	0·1006	34·6	31·6
130	9·58	0·377	72·1	0·1118	33·7	30·8
140	10·1	0·398	80·1	0·1242	32·7	29·9
150	10·5	0·413	86·6	0·1342	32·0	29·3
175	11·7	0·461	108	0·1674	30·5	27·9
200	12·8	0·504	129	0·1999	29·1	26·6
225	13·8	0·543	150	0·2325	28·0	25·6

Current in amperes.	Diameter.		Cross-section.		Voltage drop.	
	Mm.	In.	Sq. mm.	Sq. in.	Volts per km	Volts per 1000 yd.
250	14·9	0·587	174	0·2697	26·8	24·5
275	15·8	0·622	196	0·3037	26·2	23·9
300	16·8	0·661	222	0·3441	25·3	23·1
325	17·7	0·697	246	0·3813	24·7	22·6
350	18·6	0·732	272	0·4216	24·1	22·0
375	19·5	0·768	299	0·4635	23·5	21·5
400	20·3	0·799	324	0·5022	23·1	21·1
425	21·1	0·831	350	0·5425	22·7	20·8
450	22·0	0·866	380	0·5890	22·1	20·2
475	22·8	0·898	408	0·6324	21·8	19·9
500	23·6	0·929	437	0·6774	21·4	19·6
550	25·1	0·988	495	0·7673	20·8	19·0
600	26·6	1·047	556	0·8618	20·2	18·5
700	29·5	1·135	683	1·059	19·2	17·5
800	32·3	1·272	819	1·269	18·3	16·7
900	34·8	1·370	951	1·474	17·7	16·2
1000	37·3	1·402	1093	1·684	17·1	15·6

THE CHARGE AS MATERIAL FOR FURNACE CONSTRUCTION

The idea already intimated in previous pages that the charge of the electric furnaces should also form the furnace constructional material is in itself, not new. And this it is all the better able to do, as here the source of heat is situated in the interior of the charge itself, and the heat has not first of all to penetrate through the furnace walls. In the operation of other metallurgical furnaces cooled metal bodies have for some time past been used just where the smelting furnace, formerly exclusively constructed of masonry, mostly suffered through thermal and chemical actions. The products of fusion solidified on those surface walls of the cooled bodies which face the interior of the furnace, and so formed the most refractory of protecting walls from the charge itself. In the electric furnace this must be made use of to the fullest extent. The portion of the charge which is kept in a cold and solidified condition on the inner wall surfaces by the air- or water-cooled walling of the vessel holding the charge has two duties to fulfil—

It has to prevent the contamination of the fused products with foreign matter by permitting the melt only to come into contact with substances from which it has been obtained, and to maintain the opposite electrodes electrically insulated from one another as well as from the walls of the vessel during the operation. The latter task is all the more important, as the *number of substances which are conductive under the working conditions of the electric furnace is far greater* than one was formerly accustomed to suppose. The *Héroult* process afforded us one of the first proofs of this, and during the past few years much valuable light has been thrown on the subject by the investigations of *Nernst*. Nearly all fusible metal compounds which when cold and in the solid state, *i.e.* in the absence of a flux, can from a practical standpoint be regarded as non-conductors commence to conduct for a moderate rise in the temperature (400° C. to 500° C.). Fortunately, the conductivity of the fused substances which enter into consideration for metallurgical purposes remains low enough for the desired heating to be attained with current densities which can be conveniently worked with carbon which, compared with other simple conductors, is a comparatively bad conductor. In other words, *the current density in the carbon necessary for these operations may be greater than the current density of the fusion which functions as the resistance.* Thus, the cross-section of the carbon pole may be smaller than that of the electrolyte. In this way we are given the possibility, where the working requirements may seem to render it desirable, of hanging the carbon pole in a vessel or liquid column of sufficient width that enough space remains over between the carbon and the walls of the containing vessels for feeding in the charge and for the escape of the disengaged gases, as will be seen from the illustration (Fig. 6) of the furnace for the electrolysis of aluminium oxide.

If *at first it was felt to be an inconvenience that the fusible electrolytes did not conduct when cold*, because to start the electrolysis the furnace had first of all to be raised to the necessary temperature either by arc heating or the introduction of resistance into the circuit, to-day the above-mentioned property of such substances cannot be estimated too highly in connection with the construction of electric furnaces. As already expressed above, and as we have seen from many examples in the first portion of this book, when cooled they themselves form the insulating material indispensable in an electric furnace, and are the least disturbing on account of their chemical properties.

That no reference is made to the application of the furnace charge for the inner lining of the furnace requires no special confirmation. Further, the thin linings of the vessels containing the charging material and of the furnace shells can be protected from melting by water or air cooling. Such linings consist of the well-known refractory materials—fire-clay, quartz, magnesite, dolomite or limestone, bricks of charcoal under certain circumstances, or coal-dust admixed with these materials.

CHAPTER IX

THE APPLICATION OF ELECTRIC FURNACES

Advantages.—In its application, the electric furnace forms a highly valuable addition to fire-grate furnaces. Up to the present it has undertaken the problems which are most difficult or impossible of solution with the latter. Not only *are we less dependent upon furnace constructional materials than in the application of combustion furnaces, but we can concentrate any quantity of heat in the smallest space, produce any temperature up to about* 3500° C., *and work in any kind of gas or vapour under pressure, or in vacuo, or, more correctly, in an attenuated atmosphere.* Yet another advantage should be emphasized—*the rapidity of the heat production.* This can very decisively influence the success of an experiment. For instance, in the year 1888 I succeeded in reducing the volatile molybdic acids with carbon in the apparatus depicted in Fig. 80 on p. 62. To perform the same experiment in an externally heated crucible, placed in a draught or regenerative gas furnace, all the molybdic oxide would have escaped long before the interior of the crucible had reached the temperature of reduction.

Resistance Heating.—Not on theoretical grounds only should we acknowledge resistance heating as the most perfect of the electrical heating methods. The heat is directly generated from electrical energy by aid of the substance to be heated. It is produced, as a rule, just where it is required for chemical changes, *i.e.* for the production of the durable and easily transferable chemical energy. The temperature need not be raised above that required by the smelting process under consideration, and as regards heat distribution, even in comparatively large masses of material, only minor difficulties are met with. Even in regard to the highest attainable temperature the restrictions imposed are no different from those with arc heating. For, by using the highest ohmic resistance material, we can subject it to a current density with which its temperature of volatilization is reached in the shortest space of time; and the temperature of 3500° C. also forms the limit of arc heating. In both methods the top temperature limit is defined by the evaporating carbon.

Arc Heating.—Nevertheless, there are cases enough where arc heating will practically prove of greater utility, even with conducting material present, than resistance heating. Such cases are—quick production of the highest temperature limits where resistance heating is excluded, for instance, on account of any undesirable action of carbon which must be considered for the resistance material when the highest temperature is required; and smelting processes, in which for accelerating the reactions, at least locally, the maximum conversion of energy is required. As an example, let us consider the *Héroult* electrical steel furnace at a moment when both electrodes form arcs above the slag layer and have their highest current density. When the current flows from the electrodes, an air, or other gas or vapour, layer between the electrodes and the slag forms the heat-producing resistance in the circuit; we thus at once have the possibility of arc heating. Taking the area of the bottom surface of the electrodes to be 1000 sq. cm. (approximately), with a current of 3000 amperes, in the space of a few millimetres between the electrode and the slag, assuming only 40 to 45 volts for the arc voltage ($Q = 0{\cdot}24 \times E \times C \times t$), a quantity of heat is produced of 30 kg.-calories per second in round numbers, *i.e.* more than 100,000 kg.-calories per hour at the foot of each electrode. The slag layer between the electrode and the metal forms the second resistance, for this layer is kept so thin that the main part of the current will not go direct from one electrode to the other, but flows from the first electrode to the metal, and then at the other end of the bath from the metal through the slag back to the second electrode. The quantity of heat that is produced in these two slag layers between the electrodes and the metal naturally depends on the thickness of the layer and its ever-increasing conductivity. Reckoning a potential difference of 10 volts here, a further quantity of 26,000 kg.-calories would be hourly added in each of these slag layers, quite apart from the quantity of heat produced in the iron itself.

Superiority of the Electric Furnace.—It is at once apparent from this, and it is also easy to establish by actual measurements, that a much larger proportion of the heat generated electrically in this way is transferred to the furnace charge than is possible in combustion furnaces, in which the heat carrier consists of a large volume of gas. In spite of the Siemens regenerative furnaces for smelting work with crucibles and on the Martin process, the method of heat transference by the individual hot gas molecules to the metal bath is very faulty. In the electric furnace, especially that of Héroult, the main portion of the heat is generated in the space intervening between the electrode and the slag, only amounting to a few millimetres in width. The bottom surface of the electrode is at the temperature of volatilization of the carbon. In this way a very considerable amount of heat passes to the smelting charge by radiation and by means of carbon vapour, which has a temperature exceeding 3000° C. The carbon vapour, starting from the electrode, is continually impinging on the upper surface of the slag, and is at the same time being eagerly consumed by the slag oxides. The fusion materials, slag and metal, themselves form additional producers of heat; they also constitute resistances in the current path, and, as we have just seen, the slag layer especially converts considerable quantities of electricity into heat. And above the slag layer, the position of the greatest production of heat, we have a badly conducting gas layer, which remains a long time in the furnace; and beneath the thin layer of slag there is also a highly conducting deep mass of metal, capable of absorbing large quantities of heat. It is easily seen that it is not only the distribution of the heat that proceeds from the main course of the current through the slag. The most energetic reaction takes place in this region between the superheated particles of the slag, the electrodes, and the constituents of the metal bath, and what is by no means of minor importance for the rapidity of the work, and thus for the total success of the operation, all these circumstances—the high current density; the enormous heat production in a narrow pass, I might almost say; the rapidity of the chemical reactions; and the changes of concentration of the important constituents of the slag and the metal (carbon, silicon, etc.)—*produce such an energetic, mechanical flow of the liquid particles* in the slag and metal that the desired changes are influenced in the most favourable manner possible.

CHAPTER X

FURNACE OUTPUTS

THE number of chemical products manufactured in the larger electrical furnaces which have been in continuous operation on a practical scale, or have at least been working experimentally for some time, is still by no means large; and, in consequence, the amount of actual working data available is also somewhat scanty.

The Aluminium Industry.—Besides the many contradictory reports which have appeared in connection with the aluminium industry, the following more reliable figures are now known :—

1. The British Aluminium Works, according to Wallace, use a bath as follows :—

Internal length	1500 mm. (4 ft. 11 in.)
„ width	750 mm. (2 ft. 5½ in.)
Current	8000 amperes
Current density (app.) . . .	7000 amp. per sq. m. (4·52 amp. per sq. in.)

2. An aluminium bath according to Winteler :—

Internal length	1050 mm. (3 ft. 5⅜ in.)
„ width	550 mm. (1 ft. 9⅝ in.)
Current	3200 amperes
Current density (app.) . . .	6500 amp. per sq. m. (4·2 amp. per sq. in.)

The published accounts of the E.M.F. required for the electrolysis of aluminium oxide differ widely. The most favourable reports give from 3 to 5 volts, others from 9 to 10 volts, as the bath voltages. Taking into account that the pressure of dissociation of aluminium (calculated from the formula $E = \frac{H}{n \times 0{\cdot}24 \times 96{,}537}$) amounts to about 2·8 volts, it is not improbable that with the bath working under favourable conditions a voltage of approximately 5 volts between the poles has been temporarily obtained; this should, however, not be taken as an average working voltage. On the other hand, the potential difference of 9 volts given in other sources is too high with the low current densities of to-day; an average of about 7·5 volts with a current density of 7000 amperes per square metre (4·52 amps. per sq. in.) and a 90 per cent. yield approximately corresponds to the working conditions on the large scale. At least 1400 E.H.P. should accordingly be allowed for the production of one (metric) ton of aluminium per twenty-four hours (1422 E.H.P. per ton), which very well agrees with the figures for the power consumption which I calculated for the extraction of aluminium from experiments conducted with a machine giving only 2 E.H.P.[1] *At that time I laid down that in twenty-four hours one ought to expect approximately two-thirds of a kilogramme* ($1\frac{1}{2}$ *lb. app.*) *of aluminium per E.H.P.; consequently, for* 1000 *kg.* (2205 *lb*) *of aluminium per twenty-four hours slightly less than* 1500 *E.H.P. would be required, while since then the figures which have been published from actual factory practice vary between* 1400 *and* 1600 *E.H.P. for the same output.* As already mentioned above, *the more probable power consumption is* 1400 *H.P. per metric ton in twenty-four hours* (1422 *E.H.P. per ton per twenty-four hours*).

Manufacture of Carborundum.—The *Carborundum Company* and *Fitzgerald* furnish the following data on the manufacture of carborundum:—

The resistance core is composed of granulated coke and a portion of previously used, and therefore better conducting, core material. It has a diameter of approximately 530 mm. (1 ft. 9 in.) and a length of 4250 mm. to 4500 mm. (13 ft. $9\frac{1}{2}$ in. to 14 ft. 9 in.), and is contained in a heating chamber the inside dimensions of which are app. 5000 mm. long, 1800 mm. wide, and 1700 mm. high (16 ft. $4\frac{7}{8}$ in. × 5 ft. $10\frac{7}{8}$ in. × 5 ft. $5\frac{7}{8}$ in.),

[1] Already published in the second edition of Borchers' "Metallurgie" (1895).

the furnace being app. 7000 mm. (23 ft. 9½ in.) long outside.[1] The furnace produces 3150 kg. (3·1 tons) of carborundum in thirty-six hours with a power consumption of 746 k.w., corresponding to 8·5 *k.w.-hours per kg. of carborundum* (3·86 *k.w.-hours per lb. of carborundum*). The voltage drops from app. 200 volts at the start to 75 or 100 volts at the finish of the operation with a corresponding increase in the current. About 8000 to 10,000 amperes are thus passed through the core at the finish, *i.e.* the current density in the resistance is from 4 to 4·5 amperes per sq. cm. (28·5 to 29 amp. per sq. inch).

Manufacture of Calcium Carbide.—The information available up to the present will be found on p. 104. Figures relating to the dimensions of the furnaces have so far not been published.

Steel Manufacture.—From a number of trials with the *Kjellin* furnace illustrated on pp. 56 and 57, 1350 kg. (1·33 tons app.) of pig iron and scrap were melted down in about seven hours on the average, with a consumption of 160 k.w. in round figures. As a portion of the fused ore had always to be left in the furnace, the yield of the trials observed by the reporter could not be determined with accuracy. In the most favourable case *the power consumption amounted to* 0·130 *H.P.-year per* 1000 *kg. of fused steel, and in the most unfavourable to* 0·187 *H.P.-year* (0·132–0·190 *H.P.-year per ton*). As far as can be determined from the illustration of the furnace and the above figures, the mean diameter of the hearth was 1660 mm. (5 ft. 5⅜ in.) corresponding to a length of 5200 mm. (17 ft.) and to a width of about 150 to 160 mm. (6 in. to 6⅜ in.). From this it follows that at least 600 kg. (11·8 cwt.) of metal were contained per 100 mm. (4 in.) of the depth of the bath. With a total metal contents averaging 2000 kg. (1·96 tons) the thickness of the layer of fused metal was a little over 300 mm. (12 in. nearly).

The *Héroult* furnace at La Praz worked with alternating current of about 3000 amperes at a pressure of 110 volts. According to the quality of the steel produced and the nature of the raw materials, 2300–3500 H.P.-hours were used with a yield of slightly less than 2·5 metric tons (2341 kg.), so that 0·111 to 0·17 H.P.-years are required per metric ton of steel (0·1128 to 0·1727 H.P.-years per ton).

[1] Cp. pp. 63 *et seq.*

With the latest construction of the electric furnace designed by *Stassano* very favourable results are reported to have been achieved, under assumptions, however, which rarely obtain together at one place.

These suppositions are:—

Very pure iron ore (iron glance containing about 93 per cent. Fe_2O_3).

Coke at not more than 45 lire per ton (metric) (approximately 35*s.* 10*d.* per ton).

Utilization of the waste gases.

Cost of the electrodes, not more than 0·30 lire per kg. (1·278 pence per lb.).

Available water-power of at least 5000 H.P., the installation costs of which do not exceed 300 lire per H.P. installed (£11 15*s.* 0*d.* per H.P.).

Electric furnaces of 1000 H.P. each with a daily output capacity of 6 tons (metric), (*i.e.* 5·9 tons).

A (metric) ton of malleable iron, melted direct from the ore on the *Stassano* method, *would require* 4000 *H.P.-hours*, corresponding *to* 0·45 *H.P.-year*—plenty in comparison with the other processes.

Relative Use of Direct and Alternating Current.—That for electrical smelting processes with which electrolysis is combined only continuous current can be used, and that for exclusively electro-thermal conversions under certain conditions alternating and multiphase currents are preferable, require to-day no further explanation. The changes which are produced by electrolysis, such as occur with continuous current in the masses to be heated, may become disadvantageous for the production of heat, especially in connection with the regularity of the heat production in comparatively large furnace charges. As high a periodicity as is required in electric lighting is, of course, not necessary for electric furnace working. Very good results were obtained, for instance, by *Kjellin* with 15, and by *Héroult* with 33 cycles per second. The generator used by *Kjellin* had 24 poles, and ran at a speed of 75 revolutions per minute; *Héroult's* machine had 16 poles, but its speed was 250 revolutions per minute.

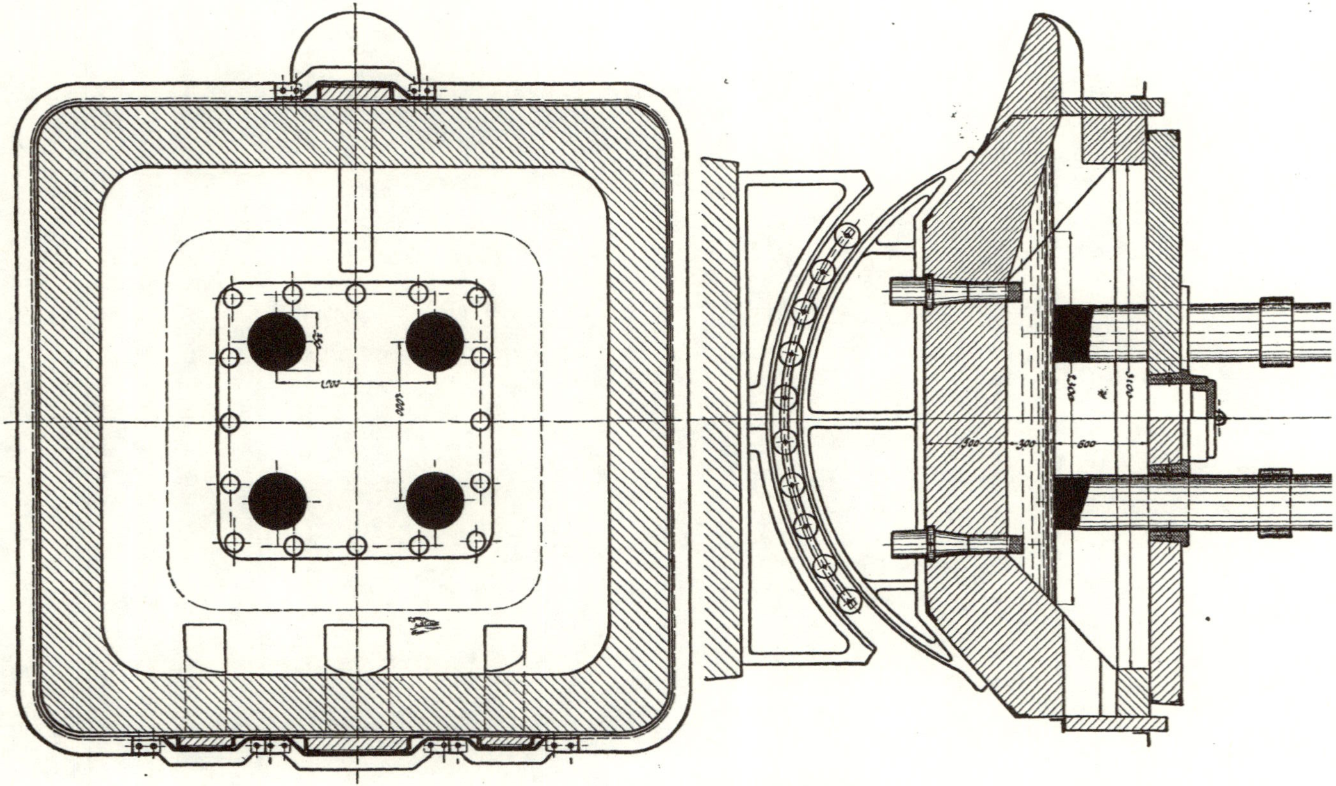

FIG. 279.—GIROD RESISTANCE-ARC STEEL FURNACE.

[*To face p.* 215.

APPENDIX

SOME RECENT DEVELOPMENTS IN ELECTRIC STEEL FURNACES

AMONG electric furnaces in which a combination of several heating systems is made use of, two specially noteworthy furnace constructions have appeared within recent years. They are the Girod resistance-arc furnace and the induction furnace of Röchling-Rodenhauser, both of which have so far been mainly applied to the manufacture of steel.

The Girod Steel Furnace.—In this furnace, constructed by Paul Girod-Ugine, arc heating is combined with direct resistance heating as in the Héroult furnace. The Girod furnace, however, of which a sectional plan and elevation are given in the accompanying folding plate, Fig. 279, excels the Héroult furnace both as regards working and simplicity of construction. The one pole of the arc, consisting of one or several carbon blocks, is introduced into the middle of the smelting hearth from above, while the other pole is formed by the slag which floats on the surface of the metal. The slag obtains electrical contact through the metal bath, which is connected to the current circuit by means of iron rods or rings inserted in the furnace from below, but near the periphery of the metal bath. These contact pieces, which are slightly cooled outside the furnace, are so proportioned that they conduct the working current without offering too large a resistance. For their conductivity they are thus nearly fully loaded, so that they ensure a good current distribution at the circumference of the bath. The ends in contact with the metal bath are slightly narrowed, so that they are kept sufficiently warm by the current to prevent undesirable loss of heat from the metal bath by conduction to the outside.

The sources of heat in the Girod furnace are thus—one or more arcs above the centre of the hearth, a layer of slag on the metal, and the metal itself. The current flows from one or several of the upper electrodes under the formation of an arc through the refining slag layer to the centre of the metal. It then radiates through the latter, flowing very symmetrically through it in all directions, at the same time heating

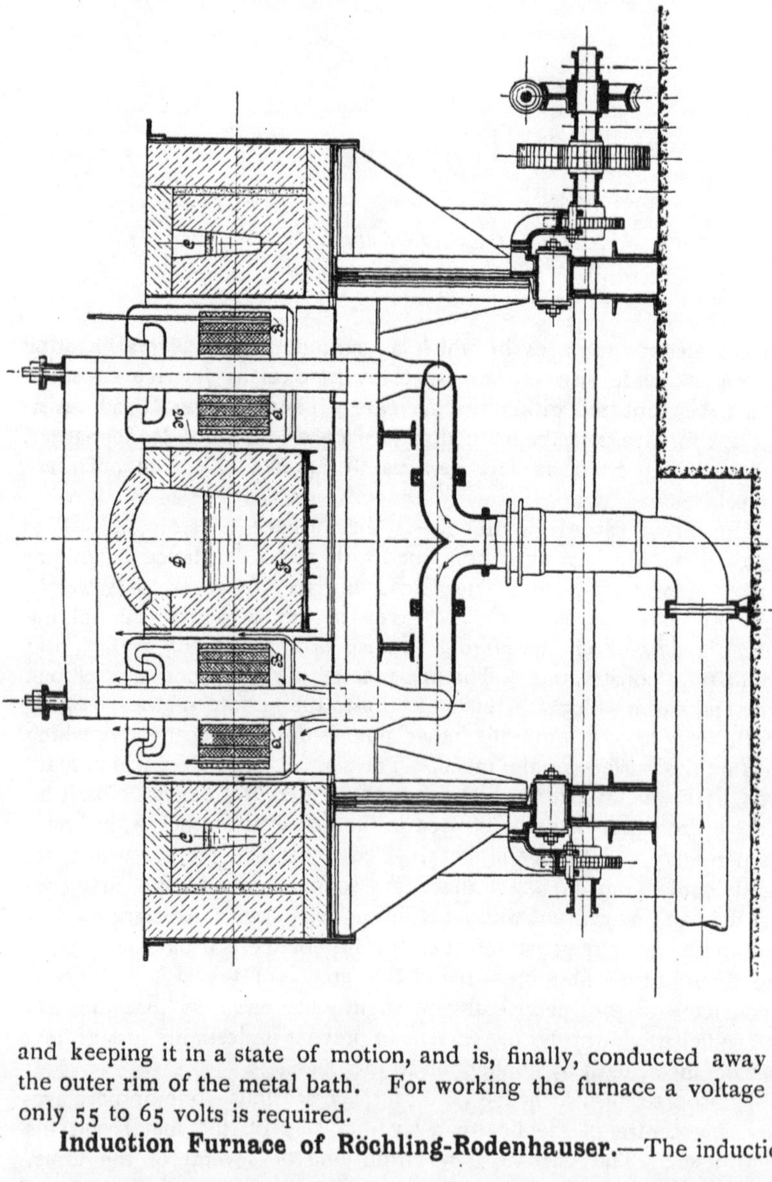

FIG. 280.

and keeping it in a state of motion, and is, finally, conducted away at the outer rim of the metal bath. For working the furnace a voltage of only 55 to 65 volts is required.

Induction Furnace of Röchling-Rodenhauser.—The induction furnaces constructed by Kjellin, Colby, and others, after the invention of Ferranti, apparently gave the best results when used as smelting furnaces, or furnaces for the production of alloys, rather than as refining furnaces. The furnace which appears to work most successfully in practice is that by Röchling-Rodenhauser. It is shown in

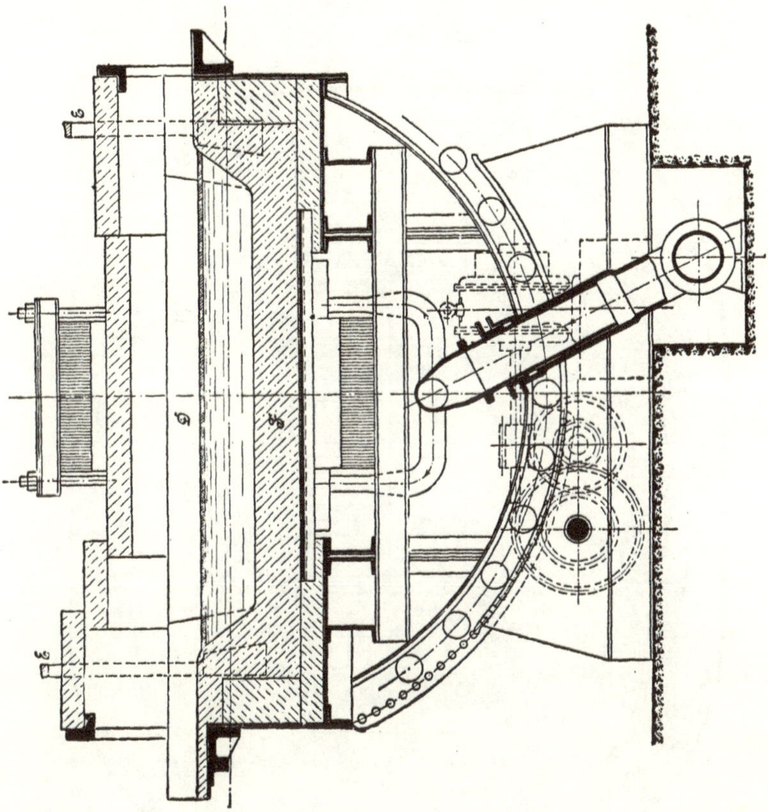

FIG. 281.

sectional, front and side elevations and plan in Figs. 280, 281, and 282 respectively.

In this furnace a portion of the bath, which forms the secondary circuit of a transformer, is also heated by a current which is sent direct through it. Both vertical limbs of the iron magnet core, H (Fig. 282), of the transformer have primary circuit windings, A, and are provided with smelting troughs, C, which run together to form a wide central hearth, D. The two protecting walls, which form the ends of this central hearth, are built up of magnesia, or dolomite, and conduct very well at the higher temperatures. Behind the protecting walls are pole-plates, E, which are connected to a secondary circuit, B, of the transformer, and from these pole-plates currents flow through the fused metal contained in the wide portion of the hearth.

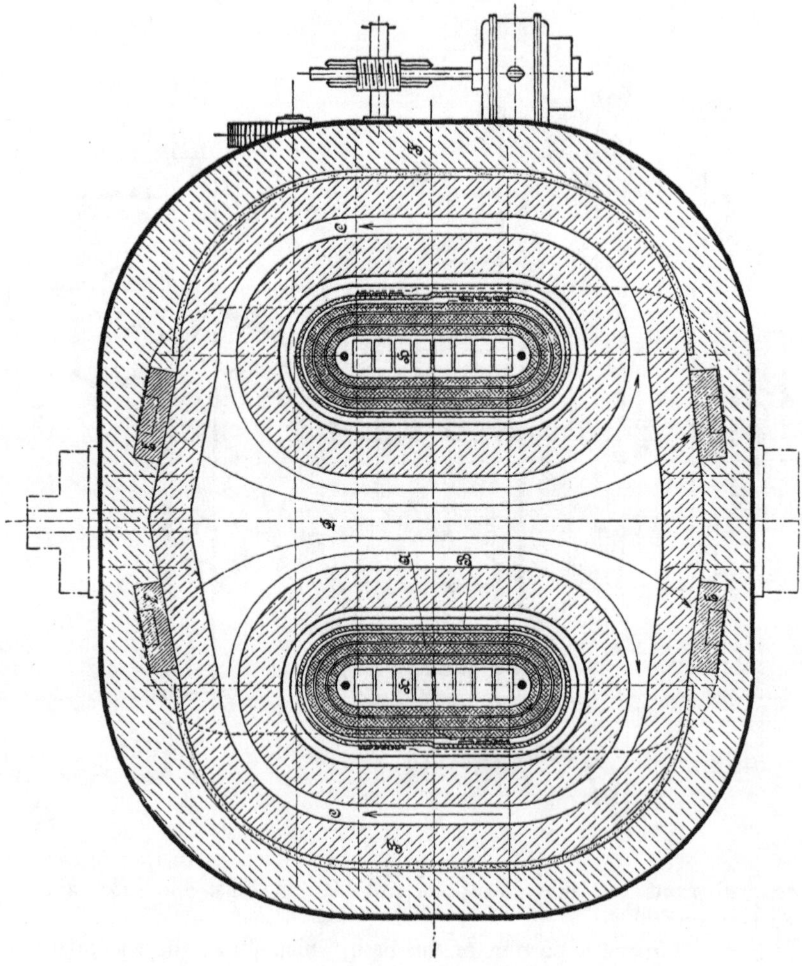

FIG. 282.

INDEX

PRINTED BY WILLIAM CLOWES AND SONS, LIMITED, LONDON AND BECCLES.

www.ingramcontent.com/pod-product-compliance
Lightning Source LLC
LaVergne TN
LVHW101442200926
848483LV00001B/20